SHIPHANDLING WITH TUGS
SECOND EDITION

SHIPHANDLING WITH TUGS

SECOND EDITION

JEFF SLESINGER

Based on a previous edition by George H. Reid

CORNELL MARITIME PRESS

Centreville, Maryland

Library of Congress Cataloging-in-Publication Data

Slesinger, Jeff.
  Shiphandling with tugs / Jeff Slesinger ; based on a previous edition by George H. Reid. -- 2nd ed.
      p. cm.
  Rev. ed of: Shiphandling with tugs / by George H. Reid. 1986.
  Includes bibliographical references and index.
  ISBN 978-0-87033-598-3 (hbk.)
  1. Ship handling. 2. Tugboats. I. Reid, George H., 1924- Shiphandling with tugs. II. Title.

  VM464.S565 2008
  623.88'232--dc22

2008025512

Manufactured in the United States of America
First edition, 1986; Second edition, 2008

In memory of Doris Slesinger

CONTENTS

PREFACE

Expertise in shiphandling and tug handling is acquired through two routes. Both are hands-on professions, and it goes without saying that one route involves the accumulation of years of practical experience. But while experience provides an essential foundation, it alone does not guarantee success. The second route is the learning process that must take place in order to translate knowledge from prior experience into new practical applications.

Successfully bringing a ship into a berth on a day when a cross wind is blowing and the current is running provides a valuable experience. However, it's one thing to remember how the ship handled on that occasion, but something else entirely to extract the kernels of wisdom from that experience and apply it to the next. No two dockings are ever identical, and it takes a shiphandler's eye, seasoned with both experience and foresight to recognize and adjust to sometimes subtle, sometimes dramatic, differences between shiphandling circumstances.

A learning process is essential to convert knowledge in the mind into practical application in the moment. One of the best means of accomplishing this is through mentoring. Professional mariners invariably speak with admiration and gratitude of the one or two individuals who passed on their "tricks of the trade" that built bridges between knowledge and practice.

Books can also be valuable in the process of learning from someone else's experience. Early in my maritime career I found such a book, George Reid's *Primer of Towing*. Its perspective was of someone on a tugboat's bridge, with a view of what did and didn't work in handling tugs and barges. In that refreshing vein, Captain Reid also wrote the first edition of *Shiphandling with Tugs*.

I have had the privilege of writing the second edition of *Shiphandling with Tugs*. My hope is to carry on Captain Reid's tradition—to provide a clear understanding of the principles of shiphandling with tugs as viewed from both the bridge of a ship and the pilothouse of a tug.

ACKNOWLEDGMENTS

I am deeply grateful to those who have helped in the production of this book. The pages that follow represent the collective expertise of the pilots, tug captains, naval architects, towing companies and training institutions that have generously contributed their time, resources and insights to this project.

Although it is impractical to name all who assisted in this endeavor I would like to specifically thank Nathan Slesinger, my son, for his work on the illustrations; Professor Edward Wellin for his editorial insight; and Liz Slesinger, my wife, for her support in this endeavor, one of the many voyages we have made together.

SHIPHANDLING WITH TUGS
SECOND EDITION

CHAPTER ONE

AN INTRODUCTION TO SHIPWORK

The purpose of this book is to provide a comprehensive guide to the safe and effective employment of tugs engaged in assisting ships to and from their berths in ports and anchorages. It is directed primarily to those who have a professional interest in the subject: pilots, docking masters, tug captains, ships' officers, and those who aspire to attain these positions. The material also may be of interest to naval architects, marine engineers, and others who may not be directly involved in the mechanics of shiphandling, but have an interest in the process.

For clarity we will define the words shipwork, shiphandling, and shiphandler as independent terms. Shipwork refers to the art of maneuvering large ships in congested, confined waterways. Shiphandling is the set of skills required to be competent in the art. Shiphandler is the person whose primary responsibility is to direct the maneuvering actions of both the ship and tugs. While recognizing the maritime roles of men and women, for the sake of simplicity, "he" replaces "he/she" throughout this book

Shipwork is not a new business. Even during the great age of sail, tugs were essential in docking and undocking ships and helping large vessels navigate narrow channels and crowded harbors.

The *Charlotte Dundas*, built in 1802, is accepted generally as the first vessel designed expressly to tow other vessels. Powered by a Watt steam engine and a paddle wheel, it worked on the Forth and Clyde Canal in Scotland. The word "tug" first entered the nautical lexicon in 1817, referring to a small towing vessel built in Dunbarton as a ship assistant.

Without the descendants of the *Charlotte Dundas* it is unlikely that the great windjammers and clipper ships would have dominated the late nineteenth and early twentieth century. Until the advent of tugs, large and small sailing ships were warped, towed by pulling boats, or kedged about whenever space was limited. Such actions would have been impossible with ships the size of the great windjammers. Big square-riggers were most at ease on the high seas. When crossing the bar or navigating in confined waters, they fared best secured to the end of a tug's tow hawser.

Today most ships are propelled by machinery. Even large sail training ships and sailing cruise ships have auxiliary engines for propulsion. The large oceangoing carriers plying today's waters are as much creatures of the high seas as the clipper ships of the past. Consequently the tug's role is as vital to maritime commerce today as it was in the days when the wind ships lay offshore waiting for a tow across the bar.

Very Large Crude Oil Carriers (VLCCs), Ultra Large Crude Oil Carriers (ULCCs), bulkers, auto carriers, and container ships are designed for efficient operation at sea. Their maneuverability and backing power are strictly secondary considerations. However, given the congestion of waterways worldwide and the ever-increasing size of the ships that call on these ports, tug assist is vital if these ships are to arrive, berth, and depart safely without delay. A competent shiphandler's qualifications must include the demonstrated ability to use tugs wisely and well.

Shipwork with tugs differs now from the days of sail. In most instances, the tug is no longer the prime mover, since the ship normally provides its own propulsion. The tug's function is to assist a vessel, rather than propel it.

By tradition, the ship's master may assume the role of shiphandler or delegate this task to a pilot or docking master. Although both the ship master and tug captain retain ultimate responsibility for their respective vessels, they do not work independently of one another. Their work toward the common goal of bringing a ship safely into or out of port is accomplished by a mutual agreement that the shiphandler will direct the actions of both the ship and tug.

This agreement has altered the historical role of the tug captain. While tug captains still are responsible for handling their own tugs, they must do so in accordance with the shiphandler's instructions. The days of a tug captain acting independently when assisting ships was left in the age of sail.

Shipwork today places the dual responsibilities of conning the ship and directing the tugs squarely on the shiphandler's shoulders, and the shiphandler should be as proficient directing the tugs as maneuvering the ship.

The essence of shipwork seems simple enough. As one professional shiphandler put it, "It is the art of moving diversely shaped objects through a fluid medium." Using tugs to assist in this task seems relatively simple. After all, a tug can only push, pull, or lie passively. Its function is to assist the ship to steer, turn, move laterally, or hold position while heaving up or letting go its mooring lines. The tug also may be required to check a ship's way (ahead, astern, or sideways), break a ship's sheer, slow its swing, or even propel it physically. In short, the tug permits a shiphandler to undertake maneuvers that otherwise would be difficult, dangerous, or impossible. To the uneducated eye it may appear that tugs are superfluous to these maneuvers, but the appearance of simplicity can be deceiving.

The experienced shiphandler has no trouble imagining the difficulty that would be encountered handling a large ship without tugs. Consider handling a large ship in ballast, in confined waters, or with a strong beam wind blowing against the enormous area of exposed freeboard. Envision docking a big ship with a strong fair current at a shaky berth, or maneuvering a deeply loaded bulk carrier with poor steering and limited backing power in close quarters. Such circumstances demand expertise from both the shiphandler and the tug captain. To function at the highest levels they must know something of each other's craft and have a mutual understanding of and respect for the skills, knowledge, and role of their counterpart.

The methods used to maneuver ships in port often differ widely, frequently depending upon local custom. In shipwork, as in other endeavors, there may be several different ways to achieve the same result. Independently of the method used, the effectiveness of the tugs engaged in this employment is determined by the knowledge and skill of those who operate and direct them.

CHAPTER TWO

PILOTING

A pilot, according to one dictionary definition, is "a person who is duly qualified and usually licensed to conduct a ship into and out of a port or in special waters." This limited definition does not speak to the special skills required of this person nor how they were acquired. Perhaps the definition should include the word practice. Practice is "the actual performance or application of knowledge as distinguished from mere possession of knowledge." Piloting is an exercise in applied knowledge.

To know the principles of shiphandling is one thing; to apply them from the bridge of a ship can be quite another. A pilot has considerable knowledge, the result of experience and study, and the skills to apply that knowledge as he stands on the bridge wing, observing the ship setting down on a dock in the midst of darkness, rain, and wind. This ability is critically important. Those that are well versed in this practice have elevated piloting to an art.

In this book the term pilot refers to more than those who hold the designated title. Pilot refers to a person who has attained the highest level of applied shiphandling expertise. In this context, docking masters, mooring masters, and qualified ships' officers may all function as pilots.

For many pilots this practice began in the small skiffs, punts, and dinghies of their youth. It was in those small vessels that they began acquiring the most important skill of a shiphandler: boat-handling feel. It was in the hours spent knocking about small harbors in small craft that many captains and pilots began developing the shiphandler's eye—the sensitivity to immediately discern waterborne motion and comprehend its cause.

It is an eye that depends greatly on the ability to pick up the visual cues of speed, direction, set, and drift through the water. It is an eye that can envision where this motion will be two or ten ship lengths in the future. The first awkward attempts to bring a small craft into a finger pier with a cross current can provide a lesson in setting up ranges to detect set and drift. Small craft can teach one how to read water—deciphering texture, color, and patterns on the water's surface that tell of wind shifts, current direction and velocity, eddies, or depth of water.

A good shiphandler engages all his senses in shiphandling and learns to listen to a vessel's sounds and vibrations. There's nothing quite like watching your vessel close in on a pier while in full reverse to refine your sense of hearing. You will never forget the sounds and vibrations of a propeller biting into good water or cavitating uselessly.

The principles of shiphandling are the same in a small skiff or a VLCC. Despite the obvious differences in scale and magnitude, expertise in both depends on the pilot using all his shiphandling senses. At any given moment, he must sense the vessel's movement and place it within the physics of maneuvering "diversely shaped objects in a fluid medium." But intuition alone is not sufficient, particularly when piloting large or underpowered vessels. A competent pilot also must think several ship-lengths ahead, clearly envisioning where his actions will take his vessel in the future. When intuitive feel and thinking ahead connect, shiphandling becomes an art.

Most pilot organizations have developed training programs designed to polish all aspects of the shiphandler's eye. The traditional elements in these programs are observation trips and mentoring by senior pilots. Simulators also have become a common form of training. Simulators have not replaced traditional methods, but are effective complementary training tools.

Simulators generally are of two types: manned model or computer-generated visual simulation. Both are excellent learning tools. The simulator platform provides an easily understood visual illustration of shiphandling principles. It allows trainees to orient to new ports or ships; practice new, routine or difficult maneuvers; and operate vessels in extreme conditions that are encountered occasionally in the field. Many facilities include sophisticated and realistic tug models or simulations. This provides a welcomed opportunity for tug operators and pilots to focus on their respective vessels' interaction and practice the coordinated maneuvers required to handle ships with tugs.

Simulators are a cost-effective way for pilots and tug captains to learn from their mistakes. As many senior pilots will attest, experience is the best teacher. The benefit of experience is learning what works and what doesn't. The greatest learning often comes from the biggest mistakes. Although the initial setup can be expensive, the cost of a simulator pales in comparison to the economic and political consequences of a true shiphandling accident. The highest price paid for a mistake in a simulator is injury to a pilot's ego.

Interestingly enough, most senior pilots prefer the manned-model simulators. This type of simulation brings one closest to the feel for the way ships handle in a variety of conditions. Sitting in a 43-foot model of a 225,000 deadweight tons (DWT) tanker, powered by a motor the size of one found in a hand drill, gives a pilot a realistic sense of the difficulty of maneuvering underpowered ships.

Prospective pilots come with a variety of qualifying experience. They may have served in apprenticeship programs or worked as deck officers aboard commercial or military vessels. Some may have worked in the towing industry, where they operated tugs doing shipwork. Many of these prospects may already be experts in the arena of their previous shiphandling experience. However, they still need to learn the shiphandling practices and waters associated with their pilotage area. Despite the level of experience, the pilot must continue to refine the skills essential to the art.

There are several different aspects of piloting. Mooring masters may be engaged to moor a vessel at a single-point mooring (SPM) or to anchor it where it can be connected to an offshore pipeline.

Others may berth a vessel alongside another ship to lighter off while either anchored or steaming. "Bar pilots" handle ships when they are crossing the bar, inbound or outbound, while "river pilots" can handle the entire transit of a river.

In some areas pilots board and disembark ships at the sea buoy and direct every phase of the operation, including docking and undocking the vessel. In other ports a docking master relieves the pilot while the vessel is docked or undocked. When tugs are used, docking masters commonly are senior tug captains familiar with using tugs in shipwork.

With the exception of the Suez and Panama canals and certain ports in Germany where pilots are in command of the ships, a pilot nominally serves in an advisory capacity. In reality, however, the pilot usually is the one who directs the maneuvers of a ship in pilotage waters.

A pilot's task is not an easy one. Pilots are frequently called upon to board unfamiliar ships (often in the middle of the night) despite high winds and restricted visibility. This task is further complicated by the idiosyncrasies of individual ships foreign to even the vessel's master. The pilot may have only a few minutes to acquire a feel for a new vessel, yet is expected to deliver a charge safely to its destination. It is an enormous responsibility. One miscalculation can be disastrous and the penalty severe. Until the seventeenth century a pilot charged with the loss of a vessel often was beheaded on the spot. An accident is no longer a capital offense, but it can cost a pilot his license and the courts can award damages against them for millions of dollars. The consequences of a serious mistake can be ruinous.

Fundamental to piloting is the ability to handle ships of different types and sizes, in varying conditions of trim and under varied conditions of wind and current. Clearly, updated knowledge of the pilotage area is essential.

An important point is that if tugs are used, the shiphandler should be fully aware of the dynamics involved so that the tugs can be employed safely and effectively. But with respect to the use of tugs,

many shiphandlers' deficiencies are evident. Tugs are often under-
or misused.

A ship sometimes maneuvers to stay in position alongside a
dock while heaving up its lines as a tug stands idly by, instead
of breasting the ship to the dock until the vessel is secured. Or
a tug lies placidly alongside a ship while a pilot is frantically
maneuvering to keep a vessel from landing too heavily alongside a
dock. A plan to use the available tugs and a few simple commands
can prevent these situations.

The advent of tractor tugs and the varied propulsion
configurations associated with them have expanded the range of
what a tug can safely and efficiently do in shipwork. However,
whether it is a modern tractor tug or an aged single-screw workhorse,
it is of paramount importance that the shiphandler understand the
tug's limits and capabilities. Although a single-screw tug rarely
is assigned to assist a ship to a berth, economics and fleet use
sometimes dictate bringing out the old workhorse.

An old joke among single-screw tug captains was the order to
"back full at 90 degrees" while the ship was moving at 3 knots. The
new version of the joke is the order to "direct pull at 90 degrees"
given to a tractor tug while the vessel forges ahead at 10 knots. Both
commands are equally impossible for their respective vessels and
highlight the pilot's lack of experience in shiphandling with tugs.

In the past it was common for pilots to pride themselves on their
ability to handle ships without assistance from tugs. Many were
excellent shiphandlers without tugs, but their performance was
average when tugs were used. The point is that shiphandling with
tugs is like any other skill, and "if you don't use it, you lose it,"
or perhaps never acquire it at all. The lesson here is clear: if the
shiphandler does not learn how to use tugs when they may not be
necessary, he is unlikely to know how to use them properly when
they are essential.

Using a tool in any undertaking requires some familiarity with
that tool. Individuals who drive cars need not be mechanics or
automotive engineers, but they must be aware of cause and effect in

the vehicle's operation. By the same token, a pilot who uses tugs to assist in shiphandling needs a reasonable degree of familiarity with the tug's functions and limitations. Some organizations now require their pilots to include time on tugs as part of their training. A candidate for a pilot's job who lacks experience on tugs should at least ride them enough to get a sense of their capabilities and limitations.

Pilots may frequently work with the same tugs and tug captains. They should be as familiar with the capabilities of both the personnel and the equipment. There are good reasons for this since personnel can affect a tug's performance. The old adage of "preferring horsepower in the wheelhouse to horsepower in the engine room," still stands. A pilot may consistently request a tug captain with whom he has a good working relationship, even if his tug is smaller and less powerful than others available.

Many towing companies have a few old, tired, and underpowered tugs they send out on jobs when their better tugs are laid up or busy. Some less scrupulous companies may add in the horsepower ratings of every motor, pump, and generator onboard to boost the listed horsepower of these older tugs to a marketable level. An older tug dispatched as a 3,000-horsepower (hp) tug, may be capable of 2,600 hp for intermittent periods at best. This is not a problem if the pilot is aware of the situation, but if not, he may discover this deficiency at the precise moment when his reputation is riding on the horsepower of that tug.

The pilot also must be alert to other factors that can affect a tug's performance. One tug master relates a story of assisting a cruise ship to dock on a very windy day. Passengers lined the rail, where the tug's line was fast, waving and smiling, unaware of the danger should the line part. The deck crew already had cleared out. The ship took a swing toward the dock and the pilot signaled for full astern. The tug operator held back from full astern out of concern for the passengers' safety. The ship landed heavily on the dock. Afterwards the pilot and tug operator had some heated words, until tug captain explained about the risk to the passengers from a parting line.

Pilots must not ignore the safety of the tug. Some shiphandlers feel that their job is to look out for the ship, while the tugs can look after themselves. Although tugs usually can, this is not what they are there to do and if the tug's captain gets "spooked" a few times the pilot in question will not get much cooperation.

Pilots should understand the dynamics of using a tug. They should know whether to place it forward, aft, midships, or fast alongside. They should also know if it is necessary to use a headline, towline, quarter line, or no line at all. Their judgment in these matters will largely determine their success in using tugs.

Ships are unlikely to decrease in size. Large deep-draft vessels will continue to call at our ports and sometimes berth at facilities built years ago to accommodate much smaller ships. These large ships, with their limited maneuverability, will require the assistance of tugs directed by expert shiphandlers. Given these circumstances, perhaps the definition of pilot should be amended to read, "One also familiar with the art of shiphandling with tugs."

CHAPTER THREE

THE TUGS

A tug captain was once asked to describe a harbor tug. He obliged by saying:

> "It's a vessel between 80 and 120 feet in length and of 750 to 4,000 horsepower. It may be single screw or twin screw. It may have a steerable nozzle, and may even have flanking rudders. However, there is one sure way you can tell if it is a harbor tug: its battle scars. The mast will be bent and the visor around the wheelhouse will be dented."

The description given by the captain was accurate enough, although a bit general. His reference to the battle scars of a bent mast and dented visor is an unfortunate, but not uncommon, result of working in close quarters situations.

Although many tugs are intended for general towing service, an increasing number are being designed as special purpose tugs. Their appearance usually indicates the service for which they are intended. Tugs designed for shipwork have narrow wheelhouses and masts that can be lowered easily to avoid damage when working under the flare of a ship's bow or overhang at its stern. They are heavily fendered, come in a variety of hull shapes, both above and below the waterline, and are highly maneuverable.

There are three general categories of tugs used in shipwork:

- Conventional
- Tractor
- Azimuthing stern drive (ASD)

Tug categories are defined by their primary means and location of propulsion. Conventional tugs have fixed propellers and rudders located aft. Tractor tugs have either cycloidal propulsion or steerable propellers located forward. Azimuthing stern drive (ASD) tugs have steerable propellers located aft (fig. 3-1).

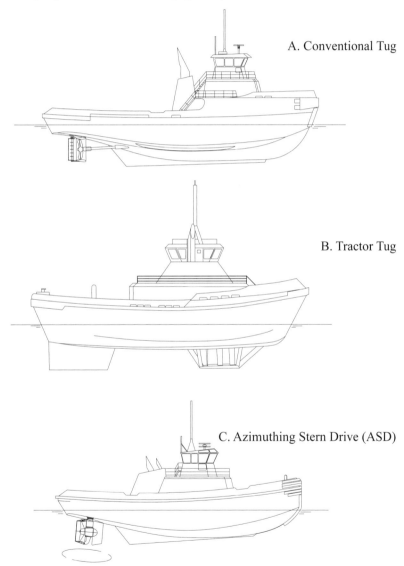

A. Conventional Tug

B. Tractor Tug

C. Azimuthing Stern Drive (ASD)

Fig. 3-1. Tug types

The fundamental goal of tugs designed for shipwork is to create a maneuverable, floating platform that efficiently turns horsepower into applied bollard pull. The key word here is applied.

One of the most efficient designs for turning horsepower strictly into bollard pull is the Navy YTB single screw tug, with its 1,800-2,000 horsepower, slow-speed engine, turning one big twelve-foot diameter wheel. Although extremely efficient, anyone who has had the "single screw blues" can attest that these tugs are not the most maneuverable. They may use up much of their available horsepower, and the operator's patience, just trying to maintain, or even get, the tug in a position where it can apply bollard pull.

Today's shiphandling tug designs have increased the efficiency of applying bollard pull by greatly improving tug maneuverability. Advances in propulsion technology, line construction, and winch design have given naval architects tools to create increasingly sophisticated and specialized maneuverable floating platforms to apply bollard pull.

The role of tugs in shipwork has expanded beyond harbors and protected waters. There are now two distinct practices: ship assist and ship escort. In ship assist, tugs are employed to assist a ship into and out of harbor berths and terminals. Ship escort is when tugs accompany a ship on its route to or from the sea.

The dedicated ship assist tug has become more powerful and compact. Ship size continues to outpace the rate at which ports build new facilities. Consequently, ship assist tugs are called on to help squeeze longer and beamier ships into narrow, constricted waterways and berths. Compared to its predecessors, the new dedicated ship assist tug has shrunk in length yet delivers more bollard pull.

Ship escort tugs provide ship escort in open roadsteads and exposed waters. Due to concerns for the environment and political pressures, mandatory ship escort routes are being pushed further and further out to sea. This requires a tug that has good sea-keeping capabilities, high stability, and winch and line technology, all of which allow the tug to apply bollard pull in open sea conditions.

This trend of designing tugs for specialized types of shipwork will continue. One example is the growing demand for tugs specifically designed to assist ships at offshore terminals. Such terminals are becoming viable alternatives for transferring dangerous liquid or bulk cargoes. As these terminals have moved further offshore, specialized ship assist tugs are being designed to function safely in offshore sea and swell conditions.

Both ship assist and ship escort tugs are becoming more specialized in design. ASD and tractor tugs, and their many design variations, are becoming more popular as the best available technology to meet the demand of both roles. Although the numbers of these types of tugs are growing, many multipurpose tugs continue to work ships in harbors worldwide. It is not uncommon to see tugs built for different services working side by side while docking ships.

Tugs assisting a ship are likely to be a diverse lot and probably will have different handling qualities and capabilities. A good shiphandler would never consider taking command of a ship without a fundamental understanding of how hull shape, size, propulsion point, and steering equipment affect the ship's maneuverability; and he should have the same understanding of tugs.

To understand a tug's maneuvering capability and limits, a pilot must use his shiphandler's eye. He must construct a mental picture of how the following elements interrelate within different tug designs:

- Towing point (TP)
- Propulsion point (PP)
- Propulsion and Steering
- Maneuvering lever (ML)
- Hull shape
- Superstructure and fendering

TOWING POINT

The towing point is visible to the naked eye. When a tug is "on the towline," it is the last physical point on the tug that fairleads its line to the ship. When a tug is pushing, it is the point of contact between the tug and the object being pushed. The towing point can be a bullnose up forward, a tow bitt aft, a staple, towing hook, or a gob line. The shiphandler must always be conscious of how his request for an application of bollard pull/push dictates the towing point on the tug (fig. 3-2).

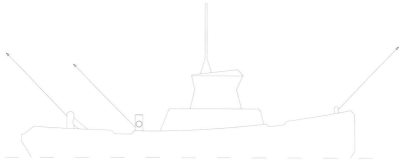

Fig. 3-2. Possible towing points.

PROPULSION POINT

The propulsion point is unseen above the waterline. It is the physical focal point of the tug's application of horsepower underwater. This point is dictated by the number of propulsion units (single, twin, or multiple), the type of propulsion (cyclodial, fixed, or steerable propeller) and the location of the propulsion units (fig. 3-3).

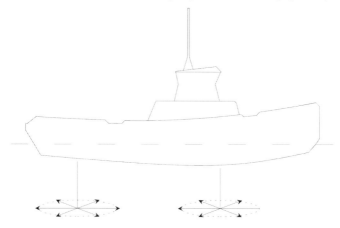

Fig. 3-3. Possible propulsion points.

PROPULSION AND STEERING

Propulsion and steering equipment determine how quickly, and to what degree a tug can change the amount and direction of thrust applied at its propulsion point. Great strides have been made in propulsion and steering technology. Cyclodial propulsion, steerable propellers, propeller nozzles, and special rudder designs have been developed to meet the need for greater propulsion efficiency and directional control. These will be discussed in greater detail in chapter 4.

MANEUVERING LEVER

The maneuvering lever (ML) is the distance between the towing point (TP) and the propulsion point (PP). It is what gives the tug leverage to swing its bow or stern around the towing point. The distance between these two points determines the efficiency of the lever; the greater the distance, the longer and more effective the lever (fig. 3-4). The tug's primary function, pushing or pulling on its towing point, is determined by the relative orientation of propulsion point to towing point. This is a key factor in distinguishing different tug design categories.

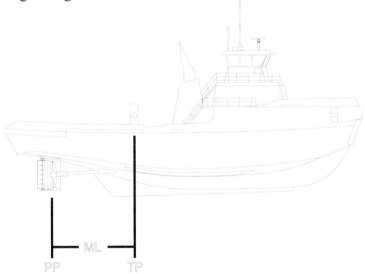

Fig. 3-4. Conventional tug maneuvering lever.

HULL SHAPE

The shiphandler must recognize that a category of tug may look similar to another above the waterline, but beneath the waterline may be very different. That difference plays a large role in a tug's ability to maneuver and apply bollard pull smartly. A tug's hull shape and appendages (skegs, propellers, rudders, and nozzles) can help or hinder movement of its maneuvering lever through the water. The shiphandler's eye must see both above and below the waterline.

SUPERSTRUCTURE AND FENDERING

Most modern ships carry long lengths of flair at the bow and tumble at their stern. The tug's superstructure and fendering determines where a tug can fit alongside a ship without damaging the ship or itself.

TUG DESCRIPTIONS

Naval architects have produced a myriad of tug designs by combining aspects of the fundamental design elements discussed above into packages that meet customers' demands.

The key is to recognize the features that have been incorporated or given up by a design to best achieve a tug's dedicated purpose. The shiphandler's understanding of the tug's handling qualities, capabilities, and limits in this respect is essential if the tug is to be used safely and to the best advantage. The handling qualities of various general types of tugs employed in shipwork, their capabilities and limitations are discussed below.

CONVENTIONAL TUGS

Conventional tugs are still the most common type of tug being used in shipwork worldwide. They can be single or twin screw, and are used for push-pull, alongside, or towline shipwork.

They are designed to spend most of their lives going forward, which is where their strength is. The hull, propeller, and rudder system are oriented to propel the tug forward. An important perspective for the shiphandler's eye is to see that, when moving ahead, the towing point on a conventional tug is forward of the propulsion point (fig. 3-5).

This type of tug maneuvers best and applies its strongest force when pushing on its towing point. On a towline this occurs when the conventional tug uses its propulsion point to push its towing point in the direction requested by the pilot and transfers that force to the ship through its towline.

When pushing on a ship, the conventional tug uses its propulsion point to push its towing point located at the bow into the ship (fig. 3-6).

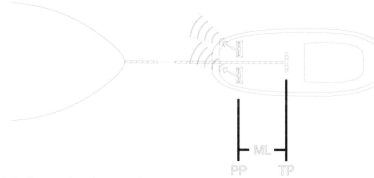

Fig. 3-5. Conventional tug towing.

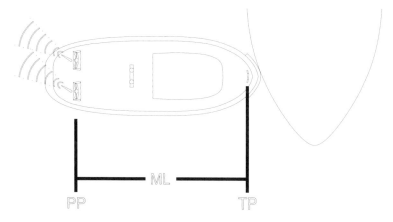

Fig. 3-6. Conventional tug pushing.

A conventional tug has two inherent weaknesses. One is a lack of astern maneuverability. This becomes apparent when the tug backs on a headline, pulling instead of pushing on its towing point. It is difficult to keep the tug in position and the tug can produce only 45 to 60 percent of its ahead bollard pull.

The other weakness is its vulnerability to capsizing while on the towline, especially if the ship has too much way on. This is due to the tug's relatively short maneuvering lever. A conventional tug is extremely powerful when the towline, towing point, and propulsion point are in line with the directional heading of the ship. However, when the towline pull begins to fall off to the side, a conventional tug must begin working its stern around to push its towing point back in line with the heading of her tow.

Under such circumstances, the length of its maneuvering lever, lateral resistance, and direction of thrust become critical. If the tug does not have enough power and leverage to rotate its stern around to push its towing point back in line, it may be in great danger. There are many terms used to describe this danger: "girt" is common in Europe, "in irons" or "tripped" are common in North America. When the tug enters into this precarious position, the pit in the operator's stomach speaks a universal language. It is the instantaneous realization of "I've lost it," and that the next sequence of events will entail bent steel, water on the decks, or capsizing. (fig. 3-7).

A *single-screw tug* is the simplest version of a conventional tug. In its most basic form it has one single flat rudder and one open propeller.

Single-screw tugs were once the principal type of tug used in shiphandling. Many of them, and the operators that have the single-screw "touch," have been around for quite a few years. There is little call for these vessels in shipwork, since more modern, maneuverable, and powerful tug designs are preferred.

The limitations of the single-screw tug are fairly obvious. It cannot maneuver as readily as a twin-screw vessel, which can turn about in its own length by working one engine ahead and the other astern. Many of the maneuvers of tugs fitted with flanking rudders, a steerable nozzle, steerable propellers, or cyclodial propulsion are beyond its capabilities.

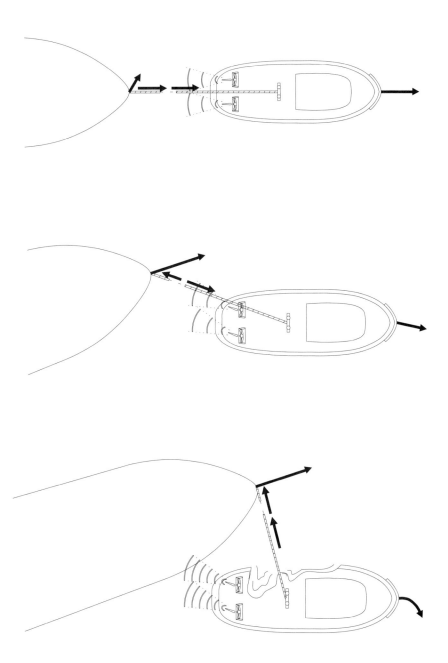

Fig. 3-7. Conventional tug tripped.

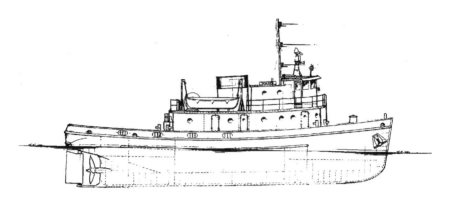

Fig. 3-8. A conventional single-screw tug. *(Courtesy of Nickum & Spaulding.)*

If a single-screw tug is required to turn in a short radius, it must "back and fill" by alternately coming ahead with the helm turned hard over and then going astern on its engine. It also turns far more readily in one direction than it will in the other because of the torque effect of the tug's propeller when it backs. In the case of a standard right-hand rotation propeller the tug will turn to starboard because while backing down, the stern will walk to port. Most single-screw tugs do not steer well when going astern, and can be difficult to control if they are required to back a long distance.

In working alongside a ship, a tug may have to press heavily against the vessel to get into position when required to push or back at right angles to the ship's side. This can be a disadvantage, especially if the ship is of light draft. Since some of the older tugs are quite narrow they are slow to turn. Their lack of beam also can make them susceptible to capsizing when working on a towline or when using a quarter line alongside, especially if the ship has too much way on.

If they are required to back for an extended period of time, it is likely that a single-screw tug will require quarter lines or stern lines (terms used interchangeably here) to maintain position alongside the ship. This is brought about because the effect of the torque of the tug's propeller causes the tug to swing out of position even if the wind, current, or the vessel's movement ahead or astern does not. (Note that the stern line keeps the tug from swinging out of position.)

Unless the average single-screw tug has been repowered, few of the older boats have more than 1,600 hp, and many of them have much less. These smaller tugs may not have adequate power to handle the large vessels that now call at our ports, although 1,600 hp usually is enough power to handle vessels in the 30,000 to 40,000 DWT range.

In spite of these shortcomings, once single-screw boats are properly made up alongside the ship they can be just as effective as other tugs, especially if they are primarily required to push. More lines may be needed to hold them in position however, especially if the chocks on the ship are poorly placed. Even today there are still shipwork applications where a single-screw tug may be preferred to twin screw or other more advanced tug designs. One example of this is in assisting submarines to and from their berths. Single-screw tugs have an advantage in assisting submarines, since their underwater configuration is less likely to damage the outer hull of the submarine (fig. 3-9). It is an odd juxtaposition, using the oldest tug design to assist one of the most modern, technologically advanced ships.

Fig. 3-9. Showing the relationship between single-screw model hull tug and twin-screw-chine hull tug alongside a submarine.

Twin-screw tugs are quite powerful. Those most commonly found in shipwork fall into the 3,000 to 7,200 hp range, and can handle the largest ships entering our harbors (fig. 3-10).

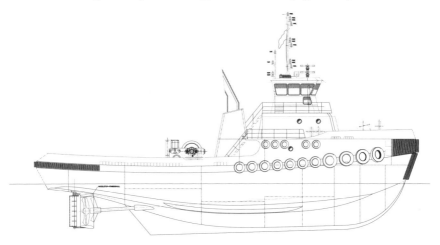

Fig. 3-10. Outboard profile of 120-foot twin-screw tug. *(Courtesy of Jensen Maritime Consultants.)*

The real advantage of a twin-screw over a single-screw tug is its maneuverability. Two engines each driving its respective propeller enable many more maneuvering options. When working a ship these tugs can shift position readily and work handily in confined spaces. They can stay in shape (maintain an angle of approximately 90 degrees) without pressing heavily against the vessel. If a ship is dead in the water and a twin-screw tug is required to back, the torque effect of the counter-rotating propellers will nullify each other, and the tug will back straight. However, if the ship is moving ahead or astern, or if there is much wind or current, the tug can maintain position at right angles to the ship only by "twin screwing," (backing on one engine while maneuvering ahead on the other with the helm turned in the appropriate direction). In this maneuvering mode backing power is less than half of full astern capability on both engines. If more backing power is required, the twin-screw tug is no different than a single screw tug—a stern line is required to hold it in shape.

Although a conventional tug with two propellers has many advantages over a single-screw tug it is not as efficient in converting horsepower to bollard pull. Horsepower-for-horsepower, single-screw propulsion is normally about 20 percent more efficient than a twin-screw installation. This is more a function of propeller size than propeller number. The single-screw tug swings a comparatively larger diameter propeller than a twin-screw tug of similar aggregate horsepower.

The shiphandler also should bear in mind that most tugs, single or twin screw, deliver much less power when backing than when coming ahead on their engines. (Refer to table 3 in chapter 4.)

Conventional tugs have been the standard workhorse in shipwork for centuries. Although they are slowly being displaced by tractor and ASD tugs, their proven design is a good shiphandling tool to pilots who know how to use them and to tug captains who know how to drive them.

TRACTOR TUGS

To the general public, *tractor tug* has become the universal term used to describe all tugs whose primary propulsion system is capable of delivering 360 degree thrust. Although this definition may be useful to the layperson, it is far too ambiguous for the needs of those in the pilothouse.

Tractor tugs derive their distinctive name from their principal method of employment. Much like a farm tractor pulling a cart, the tractor tug is designed to use its power plant primarily to pull its load.

The key difference between a tractor tug and a conventional tug lies in the orientation of propulsion point to towing point. A tractor tug's propulsion point generally is just forward of amidships. There usually are two propulsion units each capable of thrust in any direction. As in a conventional tug the towing point is aft. However, the omni-directional thrust capability and the location of the propulsion point allow the towing point to be moved farther aft than on a conventional tug. This creates a longer and more effective maneuvering lever. It is this design feature—placement of propulsion point forward of towing point—that defines a tractor tug (fig. 3-11).

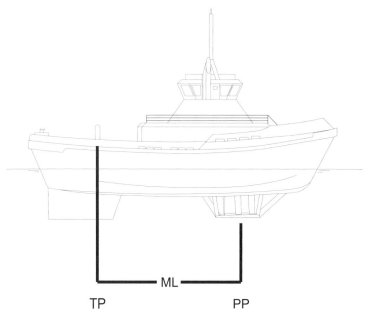

Fig. 3-11. Tractor tug maneuvering lever.

It is beneath the waterline that tractor tug designs have their defining differences. And it is here that the shiphandler can gain an understanding of the capabilities and limitations of these tugs.

If we scan the profile of a tractor tug from bow to stern, we first come upon the propulsion units. There are two basic types of propulsion systems used to power tractor tugs, the Voith-Schneider system (fig. 3-12) and the steerable propeller system (fig. 3-13). The Voith-Schneider uses a cyclodial propeller (a number of air-foil-shaped vanes that rotate about a vertical axis). The steerable propeller system employs a conventional propeller (usually fitted with a nozzle) driven by a right-angle drive from a vertical shaft, much like an outboard motor. The propeller can be turned through an arc of 360 degrees to provide both steering and reversing. Both systems require a docking plate to serve both as a guard and to provide support when dry-docked.

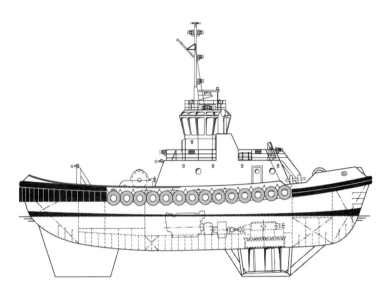

Fig. 3-12 . Voith-Schneider tractor tug. *(Courtesy of Voith Turbo Schneider Propulsion GmbH & Co.)*

Main dimensions

Length overall:	30.00 m	(98.4 ft)
Length pp:	28.40 m	(93.1 ft)
Beam moulded:	9.50 m	(31.1 ft)
Depth moulded:	3.80 m	(12.4 ft)
Designed draught:	4.60 m	(15.1 ft)

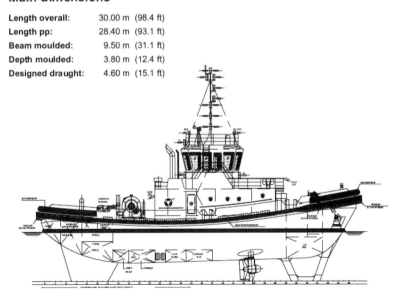

Fig. 3-13. Schottel rudder propeller tractor tug. *(Courtesy of Schottel GmbH & Co.)*

Although similar in purpose, each propulsion type has distinct strengths. In terms of turning horsepower into bollard pull, the steerable propeller system is more efficient. For a given amount of horsepower, a steerable propeller with a well-designed propeller turning in a nozzle can produce as much as one-third more bollard pull than can a cyclodial system. Steerable propeller tractor tugs tend to be deeper in the water, since the drive unit hangs down farther beneath the hull than a Voith-Schneider unit.

The Voith-Schneider system has the advantage when it comes to quickly changing the direction of thrust. A cyclodial system can initiate an almost instantaneous change of thrust with little residual thrust during the transition. However, a steerable propeller can take several seconds to rotate into a new, desired position and may produce thrust continuously over the arc of transition.

If we continue our underwater scan we next come upon the skeg. The skeg is a defining characteristic of a tractor tug. It is unique with respect to conventional tugs as it is located aft of the propulsion units. The skeg is designed to provide directional stability, lateral resistance, and hydrodynamic lift.

Early tractor tug designs required some kind of lateral plane to provide a degree of directional stability. A skeg or fin keel usually was fitted for this and to provide support when the vessel was dry-docked.

As the role of tugs expanded into ship escort, particularly high speed (eight to twelve knots) escort, the role of the skeg took on an added purpose. Skeg size and placement opened up an entirely new dimension in a tractor tug's ability to apply force to a towline. When a tractor tug has a towline up and is tailing a ship, it runs towing point and skeg first. In this aspect the skeg is located at the leading end (the stern) of the tug and almost directly underneath the towing point. A skilled tug operator can use the highly effective maneuvering lever of a tractor tug to rotate the bow around the towing point and begin presenting more of the skeg's lateral profile to the predominant water flow. This is referred to as indirect towing and if done correctly can apply far more force to a towline than engine and propeller rpm can alone (fig. 3-14). It is a powerful tool in the hands of a skilled tug master.

However, the very aspects of a skeg that make it good in ship escort applications may detract from the tug's nimbleness at slower speeds and in ship assist. This is where tractor tug skeg designs differ. Those predominately dedicated to ship escort will have long, deep, and large skegs. Those whose calling is primarily ship assist will have shorter, shallower, or smaller skegs to enhance lateral quickness.

Tractor tugs spend their lives working around ships, running forwards, backwards, and sideways. They are beamy boats to provide the necessary stability to operate in all directions with different line pulls. They have narrow pilothouses and are heavily fendered.

Tractor tugs are versatile and work equally well ahead or astern. They usually work bow first on a towline and stern first when

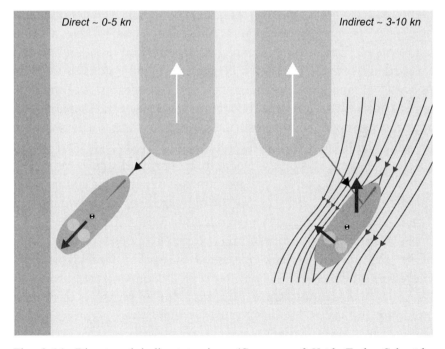

Fig. 3-14. Direct and indirect towing. *(Courtesy of Voith Turbo Schneider Propulsion GmbH & Co.)*

required to push against a ship. Their maneuverability allows them to work on a towline and then smartly shift laterally into a push/pull position alongside a ship. On the towline they are much less likely than conventional tugs to capsize, due to their propulsion point being forward of their towing point. They are much more nimble than a conventional tug performing the traditional towline, push-pull and alongside shipwork. In addition, they are capable of performing ship-assist techniques uniquely suited to tractor tugs and their close relatives, the ASD tugs.

Tractor tugs tend to be fairly operator-friendly. Generally their control systems are intuitive (they make sense without thinking), allowing a tug operator to keep his focus on driving the tug and not figuring out how to drive it.

Just as any tug, tractor tugs have their weaknesses. With the exception of specialized ship escort designs, most handle poorly in a seaway. The very characteristics that make them so handy in the harbor can detract from their seagoing capability. Their propulsion machinery is quite sophisticated and usually comes with a price tag to match. Because these expensive propulsion units are located forward, a tractor tug carries a deeper draft than a conventional or ASD tug of similar size. Since they are the deepest part of the tugs underwater profile, the propulsion units are vulnerable to damage in any grounding or encounter with underwater debris. On a horsepower to bollard pull basis the Voith-Schneider (cyclodial) system is less efficient converting horsepower into bollard pull than a propeller driven system (steerable or conventional). When lying alongside a ship delivering full thrust, a tractor tug may heel up to twenty-one degrees, increasing risk of damage to tug and ship.

A testament to a tractor tug's quick response in the hands of its operator is the lack of dented bulwarks, bent visors, and altered pilothouses visible on many of its conventional brethren. Its unblemished appearance may give the impression it can do anything. But, as in any design, tractor tugs have operating limits. Only with a clear vision of what a tractor tug looks like above and below the waterline can a shiphandler get a good sense of its capabilities and limits.

AZIMUTHING STERN DRIVE (ASD) TUGS

Azimuthing stern drive (ASD) tugs represent an evolution of multi-purpose tug design. They fuse the strong features of conventional and tractor tugs into one hull and propulsion design. From tractor tugs they take the advantage of 360 degree propulsion, which gives them a high degree of maneuverability, plus the ability to rapidly rotate around their towing point. They incorporate the efficiency of a conventional tug's propeller converting horsepower to thrust. They also take the inherent directional stability and sea-keeping qualities of a hull shape designed primarily to move bow first through the water.

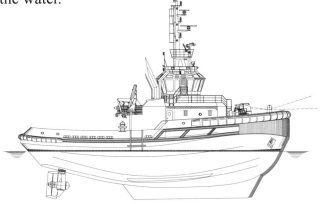

RAstar 3400 *(Courtesy of Robert Allan Ltd.)*

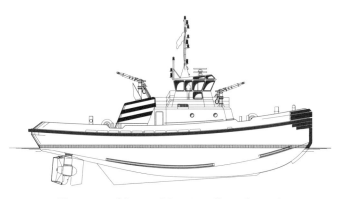

(Courtesy of Jensen Maritime Consultants.)

Fig. 3-15. Multipurpose ASD Tugs. Note different underwater profiles.

Confusion surrounds the terminology used in reference to ASD tugs. Tractor tug, reverse tractor, and Z-drive tug are all terms loosely applied to an ASD tug. The confusion arises precisely because an ASD tug can fill multiple roles. Untrained observers may use terms to define an ASD tug based on the functionality they see above the waterline. When working a ship, an ASD may look like a tractor tug and move like a tractor tug. It may perform "tractor tricks" such as indirect towing, reversing direction instantly, spinning 360 degrees in a boat length, and rapidly shifting laterally while working on a line.

Others, who are particularly observant, may notice that an ASD working a ship is "reversed;" it is working a line off its bow as opposed to its stern. Many refer to this configuration as a "reverse tractor."

When towing from a tow bitt or winch located aft of the deckhouse an ASD may look like a conventional tug. However, its quick maneuverability on the towline reflects the strength of its steerable propellers and may be referred to as a "Z-drive" tug.

This loose use of terms somewhat blurs the shiphandler's eye, obscuring the clear vision needed to use these tugs to their fullest capability. As with tractor tugs, the shiphandler needs to see above and below the waterline to fully understand how an ASD tug functions.

An ASD's defining characteristics are most obvious below the waterline. Azimuthing stern drive tugs derive their name from the type and location of their propulsion. An ASD tug always has a steerable propeller design.

Confusion arises when the term Z-drive and azimuthing are used interchangeably. Z-drive refers to an engineering design that converts the horizontal rotation of the drive shaft into the horizontal rotation of the propeller. Azimuthing refers to the 360 degree thrust functionality of these units. Azimuthing is the preferred term to ensure that functionality is more at the forefront than an engineering principle or a specific trade name.

Additional confusion is caused by the fact these azimuthing propulsion units are identical to the ones used in tractor tugs. As in a tractor tug there are two units, each independently controlled and capable of delivering thrust over 360 degrees. The location on the vessel defines their difference. On a tractor tug these units are forward. On an ASD they are well aft, situated in a position similar to conventional tug propulsion.

An ASD's towing points are forward because the azimuthing drives are aft. In an ASD tug the primary towing point for shipwork is on the fore deck. If the tug is also equipped for more conventional towing, then it will have a second winch aft. Although located in the after part of the vessel, this conventional towing point is located as far forward of the azimuthing drives as possible. (See fig. 3-16.) Above the waterline, the ASD tug may appear to be nothing more than an enhanced conventional tug. After all, conventional tugs have propulsion aft, a tow bitt, and a bullnose. Indeed, when an ASD has a towline up from its towing point aft of the deckhouse, it is functioning as an enhanced conventional tug. The ASD is moving its towing point by pushing on it and is subject to the same risk of girting, although to a lesser degree.

When an ASD uses the combination of steerable propulsion aft and a towing point at its bow, it displays its strength. With a thrust point well aft and a towing point all the way forward, an ASD has a maneuvering lever that extends the length of the vessel. This gives it tremendous leverage when working on a headline off its bow. An ASD tug takes full advantage of this lever when working in the indirect towing mode, pushing its towing point, or in direct towing when pulling on it. It is in this configuration that an ASD is operating as a "reverse tractor." Its location of towing point and thrust point are reversed to those on a tractor tug.

Although the original concept of azimuthing stern drive tugs was to provide a highly maneuverable, multipurpose tug, some ASD designs have given up the multipurpose role and focused on specialized aspects of shipwork. These designs can be identified by their appearance above waterline.

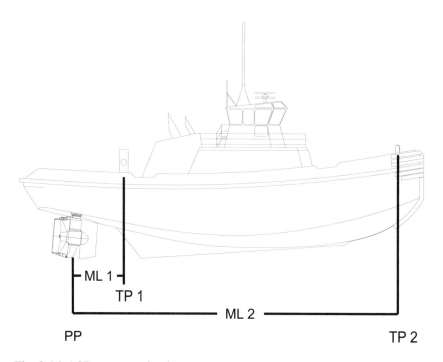

Fig. 3-16. ASD maneuvering levers.

ASD tug designs that focus on ship assist are intended to work primarily as reverse tractors. They work a headline off their forward towing point at the bow and have their deckhouse amidships or slightly aft. This is to enhance visibility forward and to keep the tug's superstructure farther away from a ship's flair or tumble when in the push/pull mode. Their largest and main winch will be forward of the deckhouse and their secondary tow bitt or tow winch will be aft of the deckhouse. Today's new generation ASD tends to have more freeboard at its stern to prevent flooding the aft deck when operating stern first at speed. Below the waterline they have a hull shape aft that enhances lateral mobility. This may take the form of a shortened keel or shallower hull aft. Fig. 3-17 is an example of a new generation ASD tug designed for shipwork.

ASD designs focused on ship escort still use the primary tow point at the bow. However the bullnose or staple is set farther back from the bow than on a typical ASD. They also ave an extended keel forward

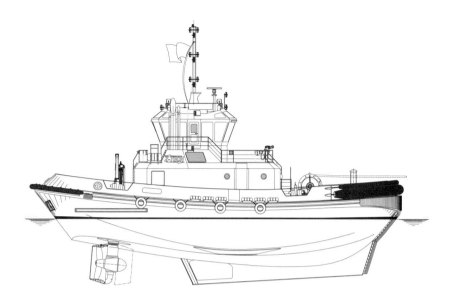

Fig. 3-17. Z-Tech 7000 *(Courtesy of Robert Allan Ltd.)*

of the towing point. This is done to enhance their hydrodynamic lift
and resistance when in the indirect towing mode. With a towing point
all the way forward at the bow an ASD would have to rely solely on
the power of the drive units to maintain the angled aspect of the hull
required to produce the lateral resistance and lift associated with
indirect towing. By positioning the staple slightly aft of the bow
and extending the fullness of the bow or keel forward of the towing
point, the ASD can enlist hydrodynamic resistance as well.

ASD tugs can be seen escorting ships in open sea, assisting them
into and out of berths and, particularly in North America, shifting
and towing barges. ASD tugs bring both functional and economic
advantages to the waterfront.

Functionally, they have good directional stability at speed
and a hull shape with good sea-keeping qualities. An ASD carries
a shallower draft than a tractor tug since the propulsion units are
located aft. The propulsion units are the most efficient design to
convert horsepower to thrust. The two independently controlled,

360-degree, steerable propellers offer a unique set of maneuvering options. In addition to functioning well in the indirect, direct, and push/pull towing modes, an ASD can slow a ship by performing a transverse arrest. In a transverse arrest both drive units are thrusting outboard, 90 degrees to the tug's keel line. The effect of the propeller wash shooting out at 90 degrees to the tug keeps the tug in a stable position and adds considerable drag and braking effect to the ship (fig. 3-18).

When ASDs move stern first towing on a headline, they share the same low risk of girting as a tractor tug. They also are capable of rapidly repositioning themselves from towing on a headline to a push/pull position alongside the ship.

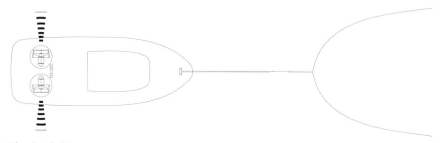

Fig. 3-18. Transverse arrest.

Economically, the initial capital outlay for an azimuthing drive unit generally is less than for a comparable cyclodial system. ASD drive units can be removed easily for maintenance, and many repairs can be completed without dry-docking the vessel. Due to their multipurpose design, they perform different types of towing services economically. In Dutch Harbor, Alaska, it is not uncommon to see an ASD tug head into the Bering Sea towing a barge upon being released after docking a container ship.

ASDs also have their shortcomings. They are less nimble than a tractor tug when moving laterally. As a light tug or when towing on a headline, they do not walk sideways as swiftly as a tractor. Changing direction of thrust is not as instantaneous as with the Voith-Schneider propulsion. The drive units typically produce thrust through the arc of transition as they are being shifted from one thrust direction to another. Their controls are not as intuitive as the Voith-Schneider and are more difficult to master.

A final word is in order with respect to the operator skill level required to drive tractor and azimuthing stern drive tugs. With all tug designs horsepower in the wheelhouse is as important as horsepower in the engine room. It is even more critical with ASD and tractor tugs. ASD and tractor tugs are exceptional vessels. In many ways they are the formula one race cars of the waterfront. They are high-performance, responsive, quick to maneuver, and extremely powerful. They come in compact hulls, bristling with new winch, line, and fendering technology. But, like a formula one race car, they need a skilled hand at the wheel. Tractor tugs and ASDs need to be driven almost all of the time. The design aspects that make them so quick and nimble in skilled hands are the same ones that rapidly can put them out of control in those of a novice. Many of the unique maneuvers performed by these vessels take them to the upper range of their safe operating limits. Staying within those limits depends on the operator's competence.

The tug designs described in this chapter are the basic and fundamental designs found at the beginning of the twenty-first century. Many others combine aspects of these designs for special purposes, such as roto-tugs, combi-tugs, and ship docking modules (SDMs). Fig. 3-19 illustrates a SDM.

Advances in tug design, equipment, and technology are happening quickly and will continue to do so. A book simply cannot keep up with these advances. However, a shiphandler can.

A shiphandler's knowledge of basic tug design principles is his tool for understanding. Regardless of new technology, a shiphandler can use the shipwork tug of the future by applying his understanding of the fundamental relationships between a tug's towing point, propulsion point, maneuvering lever, hydrodynamic lift, and lateral resistance.

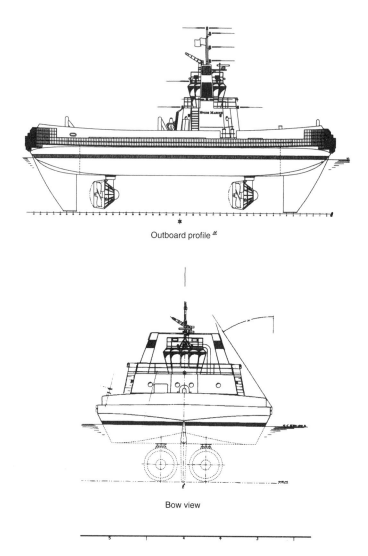

Outboard profile

Bow view

Fig. 3-19. Four-thousand hp ship docking module. *(Courtesy of Marine Inland Fabricators of Panama City, Florida.)*

The human element outweighs the mechanical in shipwork. Even a second-rate tug in the hands of a first-class operator, directed by a top-notch shiphandler, can outperform better equipment in the hands of those less competent. In the wrong hands, the most capable tug may only complicate, rather than simplify, shiphandling.

STUDY QUESTIONS

1. What are the three general categories of tugs used in shipwork?
2. What are the five key elements in tug design?
3. What is the towing point?
4. What is the propulsion point?
5. What is the maneuvering lever?
6. Describe the length and orientation of the maneuvering lever for:
   - Conventional tugs
   - Tractor tugs
   - ASD tugs
7. What causes a tug to capsize while on a tow line?
8. What advantage does a single-screw tug have over a twin-screw tug?
9. What is the principal disadvantage of a single-screw tug?
10. What advantage does a twin-screw tug have?
11. What are the disadvantages of the twin-screw tug?
12. What advantages does a tractor tug have over a conventional tug?
13. What advantages does an ASD have over a tractor tug?

CHAPTER FOUR

PROPULSION AND STEERING

Two elements vital to the movement of any vessel are its motive force (the propelling machinery), and the apparatus that controls its direction of movement (the steering gear). In the case of tugs, both merit special attention.

A tug is a floating power plant and its function is to apply this power by pushing or pulling. The hull is the platform that supports mechanical systems that are essential or incidental to this task. At the heart of these systems is the tug's steering and propulsion, the mechanism that permits the operator to control and deliver the power that the engine develops.

It may be easier to understand a tug's propulsion and steering if they are discussed separately. But one must never forget that in operation they are inextricably linked. Although their functions may be housed in separate pieces of machinery their interdependence remains. One cannot fully function without the other. The shiphandler must have a fundamental understanding of a tug's propulsion and steering machinery, and the mechanical and electronic links that tie these functions together.

All tug steering and propulsion systems combine four basic components:

1. Power source
2. Power transfer and control system
3. Hydrodynamic driver (thrust)
4. Thrust directional control mechanism

Tugs employ many steering and propulsion systems. Each has one or more characteristics that represent a technological innovation created to enhance a tug's maneuverability or bollard pull.

POWER SOURCE

Tugs were originally powered by steam engines. These engines were controlled easily and responded quickly when maneuvering ahead or astern. Engine speed could be adjusted to suit the situation, from barely turning to full speed. However, they were big, heavy power plants and had a high weight-to-horsepower ratio. The tug Hercules, built in 1907, housed one 1,000 hp, three-cylinder, triple-expansion steam engine within the confines of a 151-foot hull. Today, a tug of the same length provides a platform for two more compact and-fuel efficient diesel engines, producing a combined horsepower of 10,000 or more.

Diesel engines are now the dominant source of power aboard tugs. The evolution of the diesel engine used in towing applications has progressed from big, low-horsepower, slow-speed designs to smaller, medium- and high-speed, electronically governed engines with higher output. They are more powerful and efficient, making it possible to increase both the range and power of tugs. Compact, powerful ship-assist tug designs could not have been built without the diesel engine technology that places more horsepower in a physically smaller footprint.

A variation of the diesel engine is the diesel-electric (DE) drive. Initially developed for marine use in submarines, the system was later employed in tugs. It has proven to be an excellent method of harnessing and controlling engine output. The DE propulsion unit consists of three basic elements: the main engine or engines (usually two or more); the generators, which are engine-driven; and the. electric motors. The DC motors derive their power from the generators and drive the propeller shaft. Diesel electric drive can deliver shaft speed from dead slow to full speed ahead or astern and responds without delay to changes in speed and direction of rotation.

In addition to their maneuverability DE installations have other virtues. Engine maintenance is minimal since the diesel engine turns at a constant speed. The engine fuel consumption reflects the load on the engines. With multiple-engine installations, only the engines and generators required for propulsion at the time need to be run. Since the engines do not directly connect to a propeller shaft, DE installations allow naval architects more flexibility in engine room design. The capability to place the main engines anywhere in the engine room may free up space for extra fuel capacity, crew accommodations, or other improvements.

The cost of the DE is a drawback. Large generators (particularly the DC type) and DC motors are expensive. Any marine electrical system is vulnerable to dampness and salt in the atmosphere, which can result in costly repairs.

The silicon-controlled rectifier (SCR) system of a DE drive has proven effective in recent applications aboard small vessels, including tugs. In the SCR system, the main engine generators produce alternating current (AC), which is suitable for ship supply and other applications. The current is delivered through rectifiers, which convert it to DC for driving the propulsion motors. AC generators are cheaper than the DC versions used in older units, and since AC current is compatible with the vessel's electrical system, it can offset the cost of auxiliary generators.

Hybrid power is a new technology being developed for harbor tugs. Harbor tugs typically need high amounts of power for short periods of time. Hybrid power combines diesel engine, diesel-electric, and battery technology to meet the spectrum of horsepower demands for a typical harbor tug. Hybrid power systems also need a sophisticated power management system to mix and match power sources depending on demand.

Many harbor tugs spend a large amount of time idling, main engines on and ready to respond, but with little or no power actually being used for propulsion. The hybrid system represents cutting-edge technology, producing energy only on demand, eliminating the practice of idling main engines needlessly. This combination reduces emissions, lowers fuel consumption, and reduces noise.

POWER TRANSFER AND CONTROL SYSTEM

Whatever the power source, the end result is energy generated and transformed into a turning shaft, resulting in a rotative force. There must be a reliable means of transferring this rotative force to the tug's hydrodynamic driver (typically a propeller or the rotating vanes of cyclodial propulsion), and a mechanism for controlling the rotational direction and speed of the propeller.

Each type of transfer and control system has its own set of strengths and weaknesses, which exist within the gears, clutches, shafting, and wheelhouse controls that direct the transfer of power to thrust. The shiphandler needs to be aware of how each system's features affect reliability, efficiency, and quickness—reliability to withstand heavy use during maneuvering typical of harbor tugs; efficiency in converting horsepower to thrust; and quickness in the system's capability to alter speed and reverse the direction of the rotating propeller shaft.

The simplest form of transmission is directly connecting the tug's main engine to a shaft that has a propeller attached to the other end. The propeller then converts this rotative force into thrust. When tugs were propelled by steam engines, this arrangement worked quite well. The slow-turning steam engine matched the optimal rotational speed of the efficient, large diameter propeller swung by these tugs. However, as the diesel engine replaced steam, a new layer of complexity was needed to make the connection between power source and propeller, and reverse the direction of propeller thrust.

The direct reversing (DR) system is the simplest and least expensive system for transferring the power of a diesel engine to a propeller. It also is the oldest. But two problems occur when a diesel engine is connected directly to a propeller. The first is that the revolutions per minute (rpm) of a tug's diesel engine is higher than what is optimal for a propeller turning in water. A set of reduction gears housed in a machinery casing separate from the engine is needed to reduce these high rpms. The engine is connected to the input shaft of the reduction gear, and the propeller connected to the output shaft. Although passing through a set of gears, the engine still is directly connected to the propeller. When the engine turns, the propeller turns.

This is the root of the second problem: how to reverse the rotation of the propelling shaft. In the case of the DR system, the answer is an engine capable of running equally in either rotation. To reverse the propeller, the engine is stopped and restarted in the desired direction.

DR engines on tugs are air starting. Compressed air from a bank of tanks is admitted directly into a cylinder where a piston is under compression. This starts the engine turning and the firing cycle of the cylinders begins. The direction of rotation is determined by the starting air valve and the position of the cam shaft, which can be shifted by the operator.

This system is simple and, in less demanding applications, is satisfactory. But in harbor tug applications a diesel engine connected directly to the propeller shaft has significant disadvantages. The engine must be stopped completely and restarted when it is necessary to reverse the rotation of the propeller shaft. Its speed range is fairly limited. Slow speed generally is about half of full speed. The number of maneuvers is limited, determined by the volume of air available in the starting air tanks and the capacity of the compressor to recharge them. The DR system also can fail. The engine may not start at a crucial time or the cam shaft may fail to shift. The shaft brake may not keep the propeller from turning, and if the vessel is moving ahead or astern too fast the engine can start. These types of failures may cause the engine to start spontaneously, in the wrong direction, or not at all.

A tug operator may be unpleasantly surprised if he approaches a docked, stationary ship a little too fast and finds that there is not only no reverse, but every time he starts the engine it propels the tug faster ahead. The failure of a direct reversible has been the cause of more than a few damaged tugs, ships, and operators' reputations.

Many DR tugs have been fitted with wheelhouse controls. While this does not eliminate the fundamental problems, these controls make it easier for the captain to maneuver. He does not have to adjust to the engineer's response time, which can vary greatly. Nor does he have to suffer the shock of seeing the engineer casually leaning over the forward rail with a mug of coffee in his hand, while he, the captain, is ringing frantically for full astern!

A number of harbor tugs still are powered by DR engines and, no doubt, will continue in service as long as engine parts are available. Nevertheless, the DR system is considered obsolete, and most tugs fitted with this type of engine eventually will be replaced or repowered with more modern machinery. The DR system is less than ideal aboard tugs, which need a high degree of maneuverability. For this reason, several alternate methods for transmitting engine power to thrust have been developed.

The direct reversible and direct connect systems rely on the engine to fill three functions: generate power, transfer power, and control the speed and direction of propeller rotation. Diesel engines are at their best when called upon to generate power. The next evolution of transfer and control systems was moving the other two functions into machinery physically separate from the engine. This technology was called the controllable pitch propeller.

The moves the controlling mechanism of both the amount and reversal of propeller thrust into the propeller hub itself (fig. 4-1). CPP allows the propeller shaft to rotate in a constant direction, while allowing the blades of the propeller to rotate independently around their long axis, changing their pitch. Propeller pitch can be adjusted to give neutral pitch (no thrust) to full pitch (full thrust) in either the ahead or astern direction.

The pitch control mechanism consists of a solid shaft located inside the hollow drive shaft. The control shaft, actuated by electrical or hydraulic servomotors, engages cams on the base of the propeller blades altering pitch.

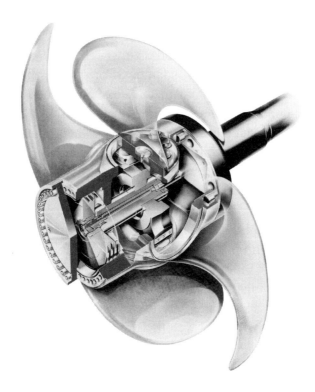

Fig. 4-1. CPP propeller. *(Courtesy of Schottel GmbH & Co.)*

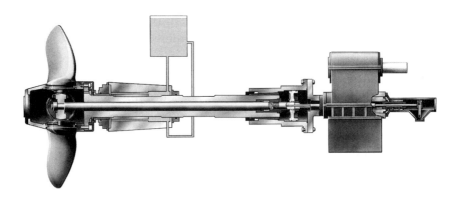

Fig. 4-2. CPP system. *(Courtesy of Schottel GmbH & Co.)*

The CPP System has three fundamental components: a reduction gear to produce a suitable shaft speed for the propeller, a hydraulic system to control pitch, and a uniquely designed propeller hub to house the mechanism (fig. 4-2).

In operation, the CPP propeller is always turning. The shaft speed can be regulated manually by the diesel engine throttle control or by a combinator control that automatically matches propeller revolutions with pitch.

The controllable pitch propeller (CPP) is proven technology. CPP systems have been in existence for some time, but have been more readily adopted by European than U.S. operators. This system has many advantages. Maneuvering with a CPP system is quite smooth. The transition from ahead to astern is seamless. Reverse thrust is accomplished by the changing pitch of the propeller blades, instead of reversing the rotation of the propeller shaft. Unlike the direct reversing system, a CPP system is not restricted by the volume of starting air.

In addition, the controllable pitch propeller can provide an appropriate pitch for many conditions of load and speed. A fixed pitch propeller is most efficient at one particular engine speed and load condition, light tug versus tug with a tow. The CPP system has the flexibility to operate across the full spectrum of a tug's load conditions. Whether barely moving, running at full speed, or applying full power on a dead pull, the operator literally dials optimal pitch.

Despite its impressive performance, the CPP system has deficits. Captains and deckhands must remain aware the propeller always is turning. For captains, this means that the tug may not coast or "dead stick" well. Even with good hull speed, the rotating blades of a propeller in neutral pitch will disrupt the water flowing by the rudders. This significantly reduces or negates rudder effectiveness. Deckhands must remain vigilant to slack lines in the water, since the constantly rotating propellers have a well-deserved reputation as "hawser suckers."

Repairing a CPP system can be expensive. The system's critical components are housed in the propeller hub, which, unfortunately, may come in contact with more than just clean water. Logs, nets, errant ropes, debris, and the occasional grounding all tend to focus their wrath on the propeller. The propeller blade may tolerate this contact, but the internal CPP mechanisms usually do not fare so well. Repair will require dry-docking and removing the tug from service. Clearly, housing the CPP's functions in a less vulnerable location has its benefits.

Reverse and reduction gears (RRG) provide this benefit by moving the reversing function out of the propeller hub and into the reduction gear itself. Additional gear sets within the gear casing convert the constant rotation of the engine's output shaft into an ahead or astern propeller rotation. The connection from the gear's input shaft to the ahead and reversing gears is made by hydraulic (oil-driven) or pneumatic (air-driven) clutches.

Historically, older forms of pneumatic clutches used heavy-duty tires to engage the forward or astern gear set. The tires had friction shoes attached to the outside perimeter. When the tire inflated, it pressed into and gripped a housing attached to the reverse or ahead gear set. Today, the most common clutch mechanism consists of a piston that pushes a series of clutch plates together connecting the engine to the desired gear set. The piston is actuated by a pressure medium. In the case of a pneumatic clutch, air is the pressure medium. In the case of a hydraulic clutch, oil is the medium.

The reverse-reduction gear is the most common form of a power transfer and control system found on conventional tugs. It is reliable, responds quickly, and can be controlled from the wheelhouse. The RRG is a relatively tough piece of machinery that can endure thousands of hours of heavy use.

While the technical details of power transfer and control systems may be of interest to an engineer, a tug operator's primary interest is knowing how the system performs under his hands. He wants to know how long does it take to shift from ahead to astern? How long does it take for the engine to come up to turns? How much power will there be when backing up? What is the limit, if any, of the number of maneuvers the system can accommodate?

The answer to these questions lies not only in the type of power transfer and control system, but in the design of the hydrodynamic driver, the propeller.

HYDRODYNAMIC DRIVERS

The purpose of the hydrodynamic driver is to convert power transferred from the engine to thrust in the water. Producing thrust in water may be simple in concept but difficult to perfect in practice. There are two types of hydrodynamic drivers employed by tugs— one in which the driver rotates around a horizontal axis (propellers), and the other in which the driver rotates around a vertical axis (Voith-Schneider cyclodial). The propeller is the staple propulsion driver in the towing industry, although the use of cyclodial systems continues to grow.

It would be relatively easy to design an efficient propeller for a vessel if it had one constant speed, displacement, and motion. But this is not the nature of tugs engaged in shipwork. Their application occurs in a dynamic, fluid environment. The propeller on a harbor tug must remain functional when the water flow is disrupted by changes in hull speed and direction, fluctuations in propeller speed, and load on the blades.

Propeller efficiency is affected by many variables: quality and velocity of water flow on the intake or discharge side of the propeller; the size and speed of the rotating propeller blades; and the number, pitch, and shape of the blades.

A large-diameter, slow-turning propeller is more efficient than a smaller propeller rotating at a higher speed. A propeller's thrust is proportional to the mass of water that the blades move and the rate at which the propeller turns. A propeller is most efficient moving a large mass of water at a slow speed. In addition, propeller efficiency is related inversely to the number of blades—as blade numbers increase, efficiency decreases. Most tug propellers have three or more blades. While four and five are less efficient, they frequently are used to dampen vibration.

Propeller pitch is the theoretical distance a propeller moves forward in one revolution. In theory, a pitch of ninety-six inches (eight feet) should move the propeller eight feet forward every time the propeller makes one revolution. The faster the propeller rotates, the faster it covers the eight-foot distance. However, the propeller is attached to a tug, so it also must move the tug eight feet. If the tug is pushing on a larger vessel the propeller will be attempting to push that vessel eight feet too. In practice the propeller never travels the full theoretical distance, since it has a certain degree of slip. Slip refers to the potential amount of power loss between propeller and vessel speed.

Tug propellers are unique in design, since they must function in a wide range of load conditions. Not only must they propel a light tug running free, but bear the heavy load of a dead push (applying full power to an immovable object). It takes a proportionate amount of horsepower to rotate the propeller at different speeds and load conditions. Engine power must be matched to pitch.

A fixed pitch propeller is most efficient at one specific speed or load condition. Selecting the right pitch is based on the tug's primary application. An ocean tug designed for good towing speed at sea will have a steeper pitch than a harbor tug designed to be strongest in a dead push. When an ocean tug is dispatched to a ship-assist job it may struggle when a pilot calls for "push full towards the dock." Too much pitch overloads the engine. If a propeller has too large a pitch, the engine does not have enough horsepower to turn the propeller adequately and the rpm level drops. The pilot may get only 75 percent of the push for which he asked.

The most common type of propeller found on tugs usually is of large diameter and fixed, but moderate, pitch. While a staple of the industry, a fixed pitch propeller has some inherent deficiencies. It does not develop equal thrust both ahead and astern. An open, fixed pitch propeller running astern will produce roughly 60 percent of its thrust ahead, and cannot have the optimal pitch for the spectrum of a tug's load and speed conditions.

Two propulsion refinements have been incorporated to address those deficits: nozzles and controllable-pitch propellers (CPP).

Nozzles (ducted propellers) consist of a circular-shaped sleeve that shrouds the propeller (fig. 4-3). The first nozzles were designed by Ludwig Kort in 1927 and have been in use since 1936. Although initially developed as a device to protect canal banks from propeller wash, they also improve propeller thrust. The nozzles are shaped like a foil and offer distinct advantages over an open wheel.

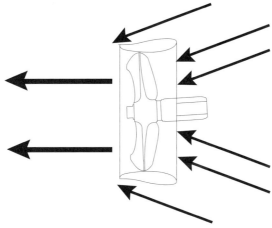

Fig. 4-3. Ducted propeller.

For a given horsepower, ducted propellers can produce 15 to 60 percent more thrust than unducted propellers. They are particularly efficient on slow-moving vessels with propellers under high load, making them well suited for tugboats.

The first tugs incorporating a nozzle were powerful when going ahead but they had drawbacks. Nozzle efficiency diminished at higher hull speeds, due to the additional drag of the nozzle. They did not steer well since the nozzle's discharge flow was not optimal for a standard rudder, reducing the rudder's effectiveness. While nozzles increased propeller efficiency they also presented considerable lateral resistance underwater, hindering maneuverability and, like an unducted propeller, their backing power was only 50 percent of ahead power.

Subsequent nozzle designs have addressed these deficiencies by refining the nozzle's foil shape, propeller blade shape, and rudder systems. Today's nozzle matched with an optimal propeller design and rudder system steers well and can produce almost 70 percent of maximum ahead thrust when backing.

The controllable pitch propeller described earlier in this chapter also addresses the performance limits of the fixed pitch propeller; optimally matching propeller pitch to load, covering the spectrum of towing conditions. However, their blade shape is not as efficient going astern and can develop only 40 to 45 percent of maximum ahead thrust.

Regardless of the type of propeller, producing thrust represents only a tug's potential capability. It is useless unless partnered with a device to control and direct it. Rudders and stern gear are called upon to fulfill this function. They are a key component of a tug's fundamental task: maneuvering into a position to apply bollard pull.

RUDDERS AND STERN GEAR

The turning force generated by a tug's rudder depends on velocity and quality of water flow and rudder shape. Rudders operate on the principle of unequal water pressures. When the rudder is turned to one side, the force of water becomes greater on the exposed or up-current side. To the casual eye it might appear to be a simple matter of the rudder deflecting the thrust of the propeller to produce a turning motion in the tug. While that is part of the rudder's purpose, it also functions as a vertical wing. As it is turned, its angle increases in relation to the direction of water flow, creating hydrodynamic lift. The combination of these two forces moves the stern of the tug in the desired direction.

Water flowing past a rudder is created by the hull moving through the water, propeller wash, or a combination of the two. Each type of water flow has its own differences in quality and velocity. A tug's rudder must be capable of functioning in all hydrodynamic conditions. Tugs particularly are dependent on propeller wash, as they do a tremendous amount of maneuvering with little hull motion through the water.

Rudder shape is a key determinant in producing hydrodynamic lift. The production of lift depends on a smooth, uninterrupted water flow over the rudder's surface. If the flow is disturbed the rudder may stall.

Rudders on tugs are oversized, since maneuverability is key to their function. Steering system designs for tugs must not only produce a powerful turning force but accommodate the strain placed on the tug's steering gear and hull structure. In addition, rudder design must be matched to the propulsion system, hull shape, and service requirements of the tug's specific design. Tugs employ a variety of rudder systems, some installed in conjunction with nozzles, to balance these required features.

Fig. 4-4. Single-screw tug with a single flat-plate rudder.

A single flat-plate rudder attached to a rudder post at its forward edge is the simplest form of a rudder system (fig. 4-4). A skeg is fitted to protect both the rudder and propeller. The rudder post fits into a bushing in an extension of the skeg. Maximum rudder angle is 35 to 40 degrees.

Although inexpensive, the single flat-plate rudder is not hydrodynamically efficient and requires heavy, large steering gear. This type of unbalanced rudder can strain the tug's steering gear, particularly when operating with sternway. All of the rudder's length is on one side of the rudder post, producing an inherent leverage that can be overcome only with powerful machinery.

For this reason, most conventional tugs are equipped with balanced or semi-balanced rudders (fig. 4-5). These have the leading edge of the rudder extending forward of the rudder post. This is done to use propeller flow efficiently and produce a mechanical advantage to the steering gear.

A spade rudder system is a balanced rudder hung independently of any skegs. It must be stoutly constructed because it has neither structural support on the bottom end nor protection from underwater debris or grounding. However, it has a good foil shape and can be located in the optimal section of the propeller wash.

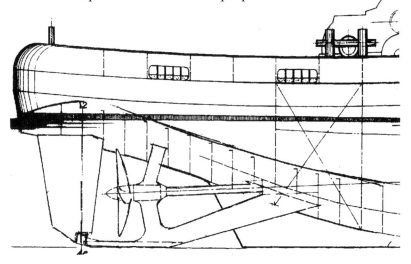

Fig. 4-5. Stern gear of a twin-screw tug with balanced rudder and skegs. *(Courtesy of Nickum & Spaulding.)*

Shutter rudders are a series of two or three high-aspect (tall and narrow) rudders mounted in conjunction with a nozzle (fig. 4-6). This type of installation permits rudder angles of up to 60°, enabling excellent maneuverability when going ahead. Towmaster, Michigan Vane, and NautiCAN are trade names associated with these types of nozzle and rudder systems.

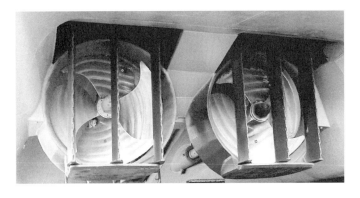

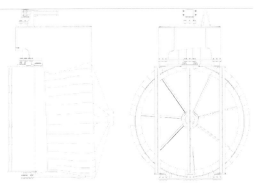

Fig. 4-6. NautiCAN nozzle and rudder system. *(Courtesy of NautiCAN Research and Development.)*

Rudder systems in which the rudder is located aft of the propeller are most effective going ahead. Water flow is generated by both hull speed and the high velocity of water coming off the propeller's discharge side. However, when the propeller is operating astern, rudder efficiency diminishes markedly. The rudder is operating on the intake side of the propeller with its reduced water velocity, and the rudder's foil shape reduces the amount of hydrodynamic lift.

One solution to this problem is placing flanking rudders ahead of the tug's propellers (fig. 4-7). Usually two are associated with each propeller, and they provide steerage when the tug maneuvers astern. They are operated by separate controls, and are kept in the amidships position during normal operation ahead. They often are installed in conjunction with nozzles.

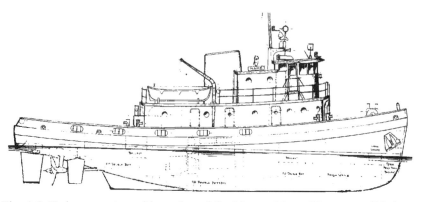

Fig. 4-7. Twin-screw tug with spade and flanking rudders. *(Courtesy of Nickum & Spaulding.)*

The steerable nozzle (sometimes referred to as a Kort rudder) provides an alternate solution (fig. 4-8). Although the position of the propeller is fixed, the nozzle is movable and turned by the tug's steering gear. Due to the increased side thrust, nozzle angles need only be 25 to 30 degrees. The movable nozzle is superior to conventional rudders in two respects: it improves the engine thrust ahead and permits the tug to steer when maneuvering astern.

The steerable nozzle does have disadvantages. It usually is slower to respond than a conventional rudder. The helm must be reversed when the engine is backed to continue to swing in a given direction. Propeller efficiency is decreased due to the increased clearance between the propeller and nozzle necessary to accommodate the nozzle's rotation. The steerable nozzle may also reduce a tug's light running speed.

Twin-screw single-rudder installations often are seen on older cruise ships. This arrangement is rare on tugs and most commonly found on vessels converted from other services. This system of steering is not very effective. It depends on the vessel's forward motion for water to flow by the rudder. Propeller wash is minimally effective because it is outboard of the rudder. While usable for hawser towing, twin-screw single rudder tugs are not handy for towing barges alongside. Since they do not have the required rudder power, they are practically useless for shipwork.

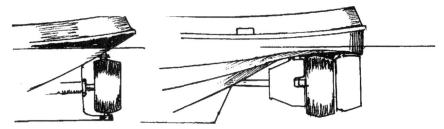

Fig. 4-8. Left: Steerable Kort nozzle. Right: Fixed Kort nozzle with flanking rudders.

Double-rudder and single-screw installations have been employed in conjunction with controllable pitch propellers. This design overcomes one of the unfavorable characteristics of the CPP system, which is a tendency to disrupt water flow to a rudder located behind it when in neutral pitch.

Conventional tugs combine the rudder and propeller systems described above in a variety of configurations. In tugs designed for shipwork the three purposes of these configurations is to provide bollard pull ahead and astern, enhance the tug's ability to steer when maneuvering astern, and maintain position when backing without having to rely upon a stern line or quarter line to stay in shape. It was noted earlier that the functions of steering and propulsion are inextricably linked and interdependent. This is best illustrated by the Voith-Schneider and Steerable Propeller Systems, where these functions are housed in a single unit.

VOITH-SCHNEIDER SYSTEM

The Voith-Schneider Propeller (VSP) functions on the principle of a controllable pitch propeller rotating around a vertical axis. On the VSP, a rotor casing ends flush with the ship's bottom. The rotor casing is in constant circular motion on a horizontal plane. Attached to the rotor casing are vertical blades (fig. 4-9). Each blade can rotate about its own vertical axis. The vertical blades perform an oscillating motion about their own axis while following the uniform rotary motion of the rotor casing. Gears and mechanical links control the pitch angle of the vertical blades at any desired point along the circumference of the rotor casing's circular motion. As a result, the same amount of thrust is generated in any direction, making this an ideal variable-pitch propeller.

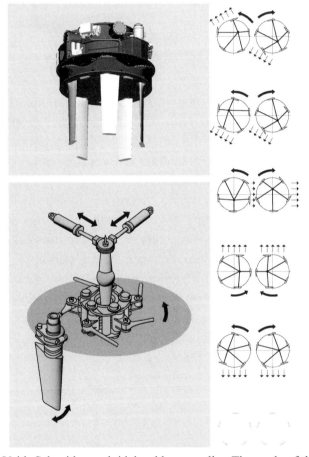

Fig. 4-9. A Voith-Schneider cycloidal rudder propeller. The angle of the vanes is changed to control steering and propulsion. *(Courtesy of Voith Turbo Schneider Propulsion GmbH & Co.)*

Tugs propelled by a VSP system have two cyclodial propulsion units mounted forward in a tractor tug configuration. Wheelhouse controls allow the operator to select rotor speed, and both longitudinal and transverse thrust. Engine rpm determines rotor speed, pitch levers control longitudinal thrust, and a wheel controls transverse thrust. Each unit has its own pitch lever to independently control longitudinal thrust. Transverse thrust control is housed in the wheel and is always applied to both units simultaneously.

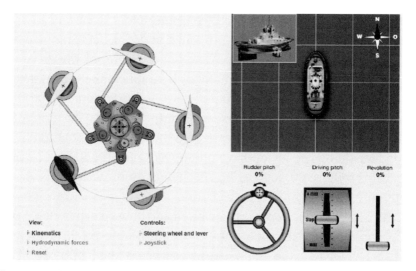

Fig. 4-10. Voith Schneider controls. *(Courtesy of Voith Turbo Schneider Propulsion GmbH & Co.)*

The control mechanisms are set so that transverse thrust overrides longitudinal thrust. When full transverse thrust is called for (wheel turned all the way to port or starboard) no longitudinal thrust is available even if the pitch levers are set at full pitch. By manipulation of the three control variables the tug operator can call for thrust in any direction.

There are several advantages to the Voith-Schneider Propeller. Like a controllable pitch propeller, changes in pitch are smooth and quick, enabling seamless changes in the amount and direction of thrust. When switching thrust directions, propeller wash disappears quickly from the initial thrust direction and transfers directly to the new desired direction.

The quick response and 360-degree thrust come at the expense of some horsepower efficiency. VSP horsepower to bollard pull ratio ahead and astern is lower than conventional or steerable propeller tugs.

STEERABLE PROPELLER SYSTEMS

Steerable Propeller Systems (SPS) come under a variety of trade names such as Z-drive, rudder propeller, and Z-peller. Engine power is transmitted through a shaft to the upper casing of the drive unit housed inside the tug. A pneumatic or hydraulic clutch regulates the connection between engine output shaft and drive unit input shaft. A series of right-angled gears and shafts converts the horizontal rotation of the engine output shaft into the horizontal rotation of the propeller shaft. The lower drive unit houses the propeller and can be rotated through 360 degrees by mechanically or electrically driven hydraulic motors (fig. 4-11).

Most SPS propellers are fixed pitch and housed in nozzles. Some SPS applications are incorporating controllable pitch propellers, particularly when the main engine serves a dual purpose. Tugs equipped with fire fighting equipment provide a good example of this application. The main engine drives both the fire pump and steerable propeller. When the situation calls for the fire-fighting pumps to be fully charged and the tug to be stationary, neutral or low pitch is used.

Most steerable propeller systems have the propeller running in only one direction. Reversing thrust requires rotating the drive unit 180 degrees instead of reversing the propeller rotation. Having the propellers rotating ahead provides the advantage of employing the most efficient aspects of nozzle and propeller design. SPS have the highest horsepower to bollard pull efficiency in both ahead and astern propulsion.

Steerable propellers can be installed in either a tractor or ASD configuration. Control of the azimuth and power for each unit is situated in one control handle. Drive units can be controlled independently or by a joystick that automatically configures the drive units for the type of maneuver selected.

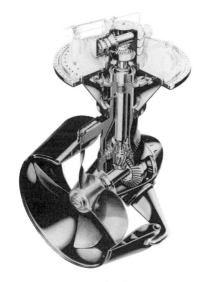

Fig. 4-11. Schottel Rudder Propeller. *(Courtesy of Schottel GmbH & Co.)*

Power versus Suitability

The standard for judging a tug's capability usually is based on its bollard pull. This is the thrust in pounds or kilograms delivered by the engine under static conditions (pulling against a dock or other fixed structure).

Bollard pull often can be predicted with fair accuracy, since tugs develop thrusts that range between 22.5 and 38 pounds per brake horsepower (bhp). This ratio usually is consistent with the tug's hull configuration, propeller type, and the presence or absence of nozzles.

Bollard pull can be affected by the tug's trim, and the number and kind of hull appendages that affect the water flow to the propeller.

Bollard pull is the common criterion for judging a tug's capability. However, other factors must be considered when evaluating a tug's suitability for shipwork, including maneuverability, stability, and backing power. A deficiency in any of these may limit a tug's usefulness even if its power is acceptable.

Fig. 4-12 gives a visual representation of the directional bollard pull efficiency of different steering and propulsion configurations.

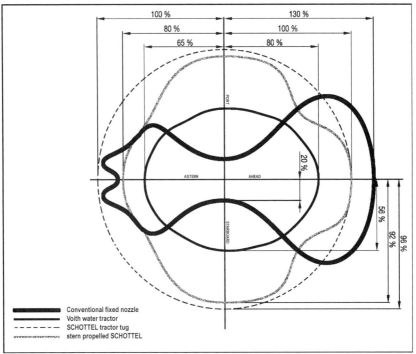

Fig. 4-12. Directional bollard pull comparison. *(Courtesy of Schottel GmbH & Co.)*

Tugs employ a variety of steering and propulsion systems. Table 4-1 illustrates the wide range of bollard pull between tugs of similar size and horsepower. Tables 4-2 and 4-3 compare tug performance associated with differing maneuvering, propulsion, and steering systems.

TUG SPECIFICATIONS

Vessel Name	Year Built/ Rebuilt	LOA (Feet)	Breadth (Feet)	Draft (Feet)	HP	Propulsion	Bollard Pull Metric Tons	Bollard Pull KIPS	Strain Gauge	Method
Valor	2007	100	40	17	6,000	Z-drive ASD	81.5	180	Y	A
Garth Foss	1994	155	46.0	18.5	8,000	Voith Tractor	79	174		A
Lindsey Foss	1993	155	46.0	18.5	8,000	Voith Tractor	79	174	Y	A
Marshall Foss	2001	98	40.0	16.0	6,250	Z-drive ASD	75	165	Y	A
Response	2002	130	45.7	19.0	7,200	Voith Tractor	70	154	Y	A
Invader	1974	136	36.5	20.0	7,200	Twin Screw	68	150	N	A
Hunter	1977	136	36.5	20.0	7,200	Twin Screw	67	147	N	A
Bulwark	1976	136	36.5	20.0	7,200	Twin Screw	65	143	N	C
Barbara Foss	1976/1993	126	34.0	14.6	5,400	Twin/ NautiCAN/ HPR	65	143	N	A
Gladiator	1975	136	36.5	20.0	7,200	Twin Screw	64	141	N	A
Jeffrey Foss	1970/1999	120	31.0	14.0	5,400	Twin/ Nautican/HPR	61	135	N	A
Protector	1996	120	41.5	19.0	5,500	Voith Tractor	55	120	Y	A
Fairwind	1975/1990	110	32.1	12.9	4,300	Twin Screw	54	118	N	A
Chief	1999	105	36.0	17.0	4,800	Voith Tractor	51	112	Y	A
Andrew Foss	1982	107	38.0	14.3	4,000	Voith Tractor	49	108	N	A
Arthur Foss	1982	107	38.0	14.3	4,000	Voith Tractor	49	108	N	A
Guide	1998	105	36.0	17.0	4,800	Voith Tractor	49	107	Y	A
Scout	1999	105	36.0	17.0	4,800	Voith Tractor	49	108	Y	A
Sea Horse	1975	126	34.0	17.0	4,860	Twin Screw	48	105	N	A
Sea Breeze	1976	126	34.0	17.0	4,860	Twin Screw	47	102	N	A
Sandra Foss	1976	111.5	31.5	11.6	2,900	Twin/Kort	42	93	N	A
Stacey Foss	1976	111.5	31.5	11.6	2,900	Twin/Kort	42	93	N	A
Daniel Foss	1978/1999	96	32.0	16.9	3,300	Z-drive ASD	41	90	N	A
Alapul	1970	105	31.1	11.4	3,000	Twin Screw	37	83	N	C
Shelley Foss	1970	90	30.0	14.2	2,400	Twin/Kort	36	79	N	A
Wedell Foss	1982	100.2	36.1	11.8	3,000	Voith Tractor	35	76	N	A
Henry Foss	1982	100.2	36.1	11.8	3,000	Voith Tractor	35	76	N	A
Drew Foss	1977	126	34.0	14.6	3,000	Twin Screw	34	75	N	A
Sidney Foss	1976	126	34.0	14.6	3,000	Twin Screw	34	75	N	A

FR = Flanking Rudder	HPR = High performance rudder	A = Actual bollard pull	C = Calculated bollard pull

Table 4-1. Puget Sound Harbor Safety Plan tug specifications. Comparative twin-screw tugboat performances gives values of thrust per bhp for different propulsion and steering systems.

Table 4-2. Summary of maneuvering systems

Type	Pros	Cons
Direct Reversing	Simplicity Low cost	Prone to failure Speed range limited Engine must be stopped to reverse rotation Number of maneuvers limited Dependent on compressed air supply
Diesel Electric (DE)	Constant engine speed Good speed range Very responsive Fuel efficient Low maintenance Suitable for large installations	Vulnerable to electrical problems High initial cost High maintenance cost
Diesel Electric-SCR (DE-SCR	Same as DE but less costly Main generators can be used for ship supply	Same as DE
Hydraulic/Pneumatic Clutches	Most common Reliable Limitless number of maneuvers Quick (four to six seconds)	Slow speed is faster than DE or CPP
Controlled-pitch Propellers	Fuel efficiency Seamless transition from ahead to astern	Complex machinery Vulnerable to underwater debris, grounding High repair cost
Voith-Schneider Propeller	360° Thrust Limitless number of maneuvers Quick response Reliability Intuitive Controls	Complex machinery Vulnerable to underwater debris, grounding High initial cost High repair cost
Steerable Propeller	360° thrust Limitless number of maneuvers Quick response Reliability	Vulnerable to underwater debris, grounding High initial cost, but less than VSP High repair cost

Table 4-3. Comparative twin-screw tugboat performance

Type of Propulsion	Controllable Pitch Propellers and Twin-spade Rudders	Controllable Pitch Propellers and Steerable Nozzle	Voith-Schneider Propellers	Steerable Propellers and Nozzles
Bollard pull to horsepower ratio	25.3-27.5 lbs/SHP (11.5- 12.5 kg/SHP)	30.8-35.2 lbs/SHP (14.0- 16.0 kg/SHP)	22.0-24..2 lbs/SHP (10.0-11.0 kg/SHP)	33.0-37.4 lbs/SHP (15.0- 17.0 kg/SHP)
Backing force (as percentage of ahead bollard pull)	30-45	30-60	90	100
Time required for an emergency stop in seconds	39	20	18	10
Time required from full ahead to full astern, in second	10	10	7	7.5
Arc over which steering force can be exerted in degrees	70	70	360	360
Steering time over full arc listed above, in seconds	15-30	16-30	15	15
Time required for 360° turn, in seconds	65-70	45-50	35-45	20-25
Turning radius in hull lengths	3-5 L	1.5-2.0 L	1.0-1.3 L	1.0-1.3 L

STUDY QUESTIONS

1. What are the four components of steering and propulsion systems?
2. What is the purpose of the power transfer and control system?
3. What effect has the diesel engine had on the design of tugs?
4. Describe the DR system of engine control.
5. What made the DR system obsolete?
6. Describe the DE system of engine control.
7. What are the advantages of DE drive?
8. What are the principal disadvantages of DE drive?
9. How does an SCR system differ from the conventional DE drive?
10. Are there any particular advantages to the SCR system?
11. Describe the CPP system.
12. What maneuvering function is housed in CPP?
13. What are the drawbacks of a CPP system?
14. How does a reverse reduction gear change propeller shaft rotation?
15. What factors determine a propeller's efficiency?
16. How do nozzles increase propulsion efficiency?
17. What are the disadvantages of nozzles?
18. What three factors determine a rudder's efficiency?
19. What is the difference between an unbalanced and balanced rudder?
20. What is a spade rudder?
21. What is the difference between a steerable nozzle and a steerable propeller?
22. Why are flanking rudders effective when operating astern?
23. Describe the Voith-Schneider cycloidal propeller.
24. What are the disadvantages of a VSP?
25. Describe a steerable propeller system.
26. What are the disadvantages of a SPS?
27. What is bollard pull?
28. Is bollard pull a fair criterion for judging a tug's serviceability for shipwork?

CHAPTER FIVE

STRUCTURAL CONSIDERATIONS

The design parameters of tugs assigned to shipwork are not solely a matter of combining high horsepower and maneuverability. The tug also must have a structural integrity that withstands the stresses placed on its hull through contact with the ship and the transfer of bollard pull from the tug to the ship. In addition the tug's shape must allow the tug to work around overhangs, bulbous bows, or other areas where the ship has limited above or below water clearance.

The nature of shipwork requires the tug to push against a ship or to pull on a line, either by backing down against a headline, or by coming ahead on a towline or springline. These routine tug actions can focus thousands of pounds of pressure or strain on various points of the tug's hull. A tug's construction and design must disperse this stress throughout the tug's structure.

Tugs engaged in shipwork regularly go alongside ships that are underway, some of which may be moving fairly fast. Although tug operators strive to make this contact in a controlled, gentle manner, human nature or Mother Nature sometimes conspires to bring the two vessels together faster and harder than planned. The tug's structure must be designed to absorb the energy of an impact.

Tugs commonly are assigned to work near the bow or stern of a vessel where the flare and overhang in these areas are likely to be pronounced. The tug's superstructure must be designed to avoid contact in these areas yet still allow the tug to align its hull at the optimal angle to pull or push.

THE HULL

Tug hulls are overbuilt by conventional standards. They must be able to withstand the frequent heavy contacts common in ship assist work without sustaining damage. The tug's hull also must be reinforced in critical areas where its propelling power is brought to bear. These areas include the bow, the stern on tractor tugs (since they push stern first), and the deck fittings that serve to fairlead or secure the tug's working lines.

The hull should be designed to have a high amount of stability, i.e., be fairly "stiff." Working lines frequently lead off to the side and exert enough heeling force that could cause a "tender" vessel to capsize.

Some multipurpose tugs carry an anchor for use in coastal towing. While it may serve a useful purpose outside the arena of shipwork, it can be a hazard when the tug is engaged in assisting ships. To reduce the risk of a fouled line or damage to the ship, the anchor should be stowed in a manner that minimizes its protrusion outside the tug's profile.

THE SUPERSTRUCTURE

Tugs that are well designed for shipwork incorporate several common features in their structures above the main deck. They have fairly low freeboard, and their deckhouses and wheelhouses are set well inboard from the rail. The bulwarks are tumbled home (angled in towards the centerline of the tug), and the mast is set up in a tabernacle for easy lowering (figs. 5-1 and 5-2). These structural design features prevent damage to the tug and the ship when the tug works under the flare of the bow or overhang of the ship's side near the stern. They also minimize the likelihood of damage when the tug is alongside the ship and both vessels are rolling a bit.

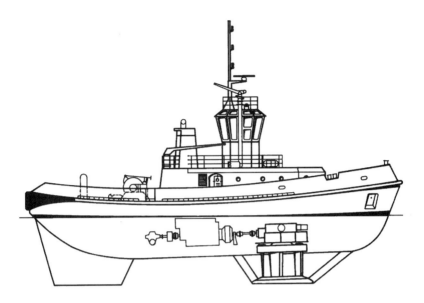

Fig. 5-1. Profile of a tractor tug. Note heavy fender installed at the stern. *(Courtesy of Voith Turbo Schneider Propulsion GmbH & Co.)*

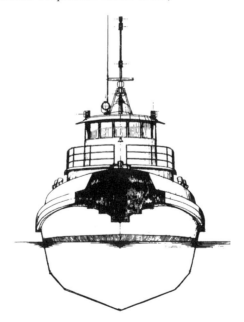

Fig. 5-2. Front view of harbor tug with multichine hull. Note that forward quarter bitts tumble in. Tug has heavy extruded side fenders.

VISIBILITY

The tug operator relies heavily on visual cues when engaged in shipwork. These cues give him essential information that allow him to gauge the motion and direction of the ship, anticipate contact points between the ship and tug, avoid obstructions created by the ship's overhangs, observe his crew on the working deck, and monitor the lead and strain on the tug's towing equipment.

For the safety of the crew, the tug, and the ship, the operator's field of view should be as unobstructed as possible. An operator's optimal horizontal view should be as close as possible to 360°, particularly for tractor and ASD tugs that are adept at working both stern and bow first. In addition to good horizontal visibility, newer wheelhouse designs enhance vertical visibility by incorporating additional small windows that face upwards. These windows are especially helpful when working alongside large vessels with high freeboard. This is a big improvement over the old-style wheelhouse, which usually had a visor that impaired the tug captain's vertical field of view (fig. 5-2). Exhaust stacks are one of the largest impairments to good visibility aboard tugs. Newer designs minimize this by moving the stacks alongside the wheelhouse (fig. 5-3).

Fig. 5-3. Tractor tug sight lines. Note location of exhaust stacks. *(Courtesy of Voith Turbo Schneider Propulsion GmbH & Co.)*

The location of the tug's maneuvering stations may also impair the tug operator's field of view. Control stations should be placed where the operator's view is unrestricted and with ready physical access to the tug's steering, engine and winch controls. This may necessitate two or more maneuvering station locations in the wheelhouse to provide adequate sight lines (fig. 5-4).

Ergonomics have become important in the design of wheelhouse layouts. The wheelhouse design of the modern ship assist tug minimizes the tug operator reaching, walking, turning, or twisting to operate the tug and see what he's doing. The maneuvering controls are laid out in a manner that makes instinctive sense to the operator regardless of which way he is facing and provide easy access to essential outside and inside visual information.

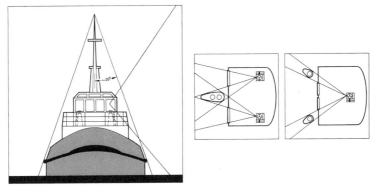

Fig. 5-4. Left: Front view of tug showing inset of superstructure and vertical sight lines. Right: Showing field of vision astern from steering stations. *(Courtesy of Voith Turbo Schneider Propulsion GmbH & Co.)*

UNDERWATER CONFIGURATION

Due to the nature of their work tugs have large single, dual, or multiple propulsion structures underneath their hull. Azimuthing drive units, cycloidal propulsion units, nozzles, and large propellers are examples of these structures. The tug's underwater configuration should accommodate these structures in a manner that ensures the propulsion units will not hit the ship if the tug is rolling while alongside.

There may be special considerations in the underwater configuration of the tug's hull. For example, single-screw tugs with molded hulls (round bottoms) are preferred for assisting submarines to dock and undock. The reason for this is that submarines have sometimes had their shell plating damaged by the inboard propeller of twin-screw tugs, and a chine hull (V bottom) has a sharp corner at the turn of the bilge that also can damage a sub's outer hull. Rolling chocks, or bilge keels, also should be dispensed with for tugs in the service of assisting submarines.

FENDERING

Purpose

Adequate fender systems are imperative on any tug doing shipwork. As one tug captain noted, "Shipwork with tugs is a contact sport, and as such requires adequate padding to protect the participating parties." Fendering is a key piece of equipment that protects the structural integrity of both ship and tug (fig. 5-5 and 5-6).

Fendering has three functions:

- absorb and dissipate energy
- enhance or reduce traction
- create space

The tug generates energy when it comes in contact with the ship, either from the impact of bringing the two vessels together or the force applied by the tug pushing on the ship. Both require a fendering system that absorbs and dissipates the energy generated in a manner that does not damage tug or ship. Fenders accomplish this by two means: the fender material absorbs energy, and the fender shape disperses it over a larger structural area than the point of contact.

Fig. 5-5. Bow fender systems.

Fender design and construction can enhance or reduce the traction or "stick" between the tug and ship. Traction can be an asset in some shipwork maneuvers. As an example, a bow fender with a high-friction coefficient can help the tug pivot out and hold position while pushing on a moving ship. In other circumstances, for example, when it is necessary for the tug to slide down the side of the ship either to reposition or to break away, a low-friction coefficient is desirable.

Fenders can compensate partially for a tug's design or structural deficiencies. In some older boats, where the deckhouse is too close to the rail, the tug's breadth can be extended by using extra-wide fenders (like tractor tire casings) or several fenders. If tugs also have wheelhouses that are too wide for shipwork, roller-type fenders often are attached to either side of the bridge to protect it from damage (fig. 5-7). The tug operator should exercise care that personnel being transferred to or from a ship are kept clear of the bridge fenders. If either vessel rolls, personnel could be crushed between fenders and ship.

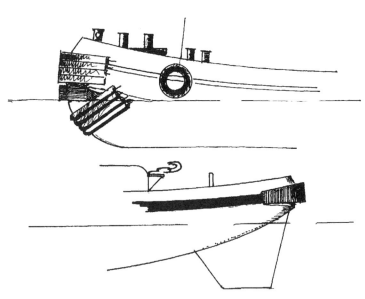

Fig. 5-6. Above: A typical installation on a harbor tug. Center: Fenders for assisting submarines. Below: Stern fender on a tractor tug.

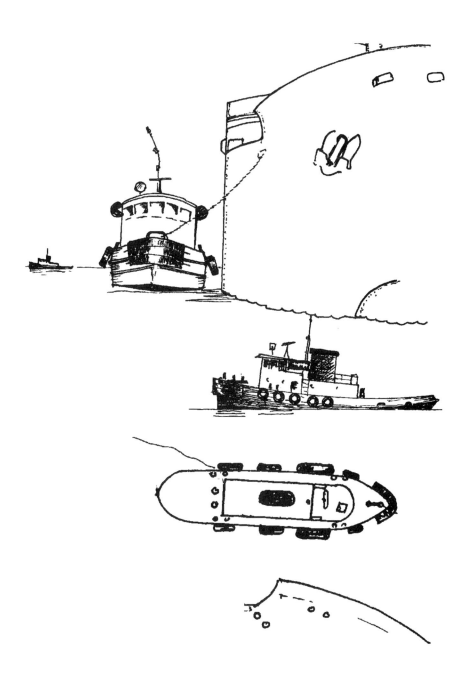

Fig. 5-7. Above and center: Wheelhouse fender arrangement on a harbor tug with a wide wheelhouse. Below: Heavy fenders can increase the tug's breadth.

Fender Materials

Wood was the original fender material. Small logs were hung over the tug's side when necessary and pulled inboard when underway. Although inexpensive, the energy absorbing capacity of wood proved no match for the evolution of heavier, more powerful tugs. Natural rope fiber became the preferred fender material. Tug crews raised the level of fender making to high art when they used old or worn towlines to construct and assemble the baggy wrinkles, turk's heads, and rope whiskers that shaped the bow and side fenders.

Rubber is the dominant material used to construct fenders today. It may come from a used airplane or heavy industrial equipment tire, extruded rubber available in a variety of lengths and shapes, or flat or looped rubber laminates cut to specific lengths and compressed onto steel rods.

Rubber fender materials also can be augmented by foam or air. Foam-filled and pneumatic fenders provide the extra energy absorption required when working in exposed, unsheltered waters.

Fender Design and Placement

Designing and placing fenders has evolved from the art of making fenders out of rope to a science. Materials are selected for a specific load deflection and energy absorption characteristics, formed into shapes designed to disperse energy throughout the tug's hull structure, and mounted at the focal points of contact between tug and ship.

There is no one-size-fits-all fendering system. Tugs engaged in shipwork typically use a combination of looped laminates, extruded fenders, and the old standby, tires, to construct vertical and horizontal fenders that suit the tug's purpose.

The objective of all fendering is to ensure that the push or contact on the ship stays within the deflection capacity of the steel plating of the ship and tug. In other words, contact without denting.

The shape of the tug's contact points is as important as the absorption factor of the fendering material. Neither ship nor tug can withstand the pounds-per-square inch loading of a high-horsepower tug focused on a small area. The load must be distributed over a larger surface. Powerful tugs have a wide radius built into their working ends for this reason. As the focal point of the tug's push against the ship, the working ends (bow and stern) must be heavily fendered (fig. 5-8).

Fig. 5-8. Combination of cylindrical and loop fendering on wide-radius bow. *(Courtesy of Schuyler Rubber Company, Inc.)*

In addition, the tug's sides have fendering to protect both ship and tug during contact that occurs as part of normal ship assist activity. This fendering frequently consists of tires or extruded rubber. The hard, slippery surface of extruded rubber makes it well suited for side fendering where less traction is desirable.

Tugs that see service working submarines must have fenders on the bow that extend below the waterline, since this is the part of the stem that comes in contact with the submarine's hull (fig. 5-6). While these underwater fenders provide necessary protection, they also may affect the tug's handling qualities. The fender's extra drag may cause the tug to steer less well than a tug without them.

DECK LAYOUT

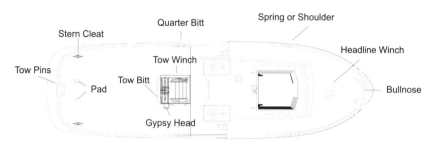

Fig. 5-9. Conventional tug deck layout.

The deck layout of a well-designed conventional tug will have a "bullnose" (a large closed chock) installed near the stem, but far enough back to avoid contact with the ship's side when the tug is pushing. These fittings are used to lead the tug's headline or springline from a position as far forward as possible. The bullnose should have an aperture large enough to permit the easy passage of the eye and splice of one or two large size working lines (fig. 5-10).

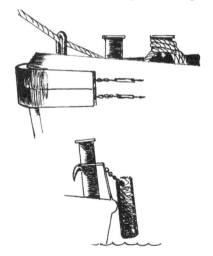

Fig. 5-10. Above: Bullnose on tug's bow. Below: Showing inset quarter bitts from the tug's side.

The lines are secured either through the use of a headline winch or belayed on a heavy H-bitt set fore and aft in the forward deck.

Both the winch and H-bitt should be stout enough to withstand the heavy line strain when the tug is backing or coming ahead with full engine rpm. Bitts used in conjunction with the newer Aramid and HMPE synthetic lines must be particularly well constructed. These lines have a high breaking strength and have been known to crush the cylindrical walls of bitts originally designed to hold traditional synthetic lines.

Forward quarter bitts are fitted on both sides of the hull, usually just forward of the deckhouse. They are set inboard so that a belayed line is not pinched or chafed when working alongside a ship or barge (fig. 5-10).

A tow winch or H-type tow bitt, set athwartship, is located aft of the deckhouse. The tow winch or tow bitt serves as the towing point and should be far enough forward of the rudders to create a good maneuvering lever between the tug's towing point and propulsion point. This distance usually is between a third and a quarter of the vessel's length forward of the stern.

After quarter bitts also are installed on both sides, usually slightly abaft the towing bitts and a capstan may be positioned to one side or the other of the tow bitts or tow winch.

The tow hook is an alternative to the H-bitt. Due to European tug's employment in towline work, a tow hook often is located both forward and aft on these tugs. Towing hooks are not often seen on American conventional tugs (fig. 5-11). The tow hook on a conventional European harbor tug usually is located almost amidships. Although this location permits a tug to maneuver well on even a short towline, it also increases the risk of the tug becoming girt or capsizing.

The main virtue of towing hooks is that they can be tripped (often by remote control) to release a hawser, even if the hawser is under heavy strain. This can be critical in averting a capsize or other emergency situation.

Fairleads and cleats frequently are installed in the afterdeck to facilitate heaving in towlines and stern lines. They also may be used to give a better lead to a stern line even though it may be belayed on a quarter bitt or the tow bitts.

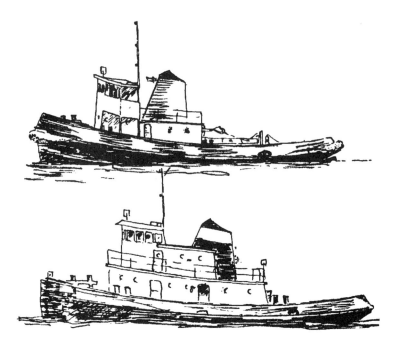

Fig. 5-11. Drawings illustrate the difference between (above) a European conventional harbor tug that has a tow hook located almost midships and (below) an American conventional harbor tug with H bitts aft, located about one-third of the tug's length from the stern.

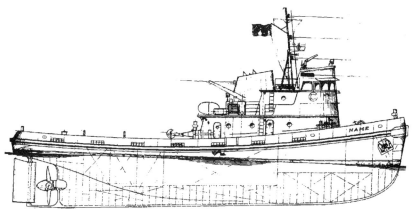

Fig. 5-12. Single-screw harbor tug with a controllable pitch propeller. The tug is designed for use in West Australia, where towline work prevails. Note the location of the towing hook and the towspans to keep the hawser from fouling fittings on the deck. *(Courtesy of Nickum & Spaulding.)*

Tractor and ASD Tugs

One major difference between the deck layout of the conventional tug versus either the tractor or ASD tug is the presence and location of a staple (fig. 5-13). The staple functions as the tug's towing point and is subject to extraordinary strain. Tractor and ASD tugs can use the hydrodynamic resistance of their hulls to place line pulls that exceed the tug's maximum bollard pull. The staple is the primary focal point of this force. It cannot be considered an appendage attached to the deck. The staple itself must be heavily constructed and mounted in a manner that solidly connects it to the tug's frames, bulkheads, and plating that create the hull's web of strength and structural integrity.

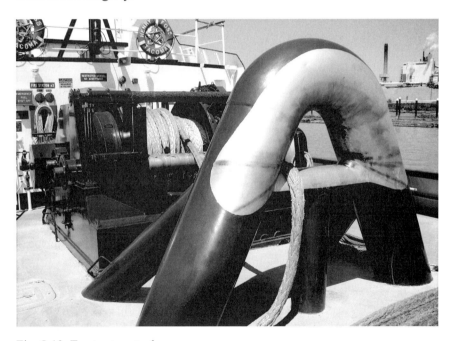

Fig. 5-13. Tractor tug staple.

ILLUMINATION

Shipwork is a twenty-four hour business and much of it takes place at night. As noted, the tug operator relies heavily on visual cues in shipwork. Tugs should have adequate lights to illuminate:

- *The working end(s) of the tug.* These are the primary contact points between tug and ship.
- *The towing point and towline.* The tug operator must be able to distinguish the direction of the line pull and whether the line is slack or under strain.
- *The deck crew's work areas.* The deck crew needs light to safely carry out their line-handling deck duties.
- *A reasonable distance ahead or behind the tug.* The tug operator should be able to see a few boat lengths ahead or behind to identify obstructions and get a visual sense of the tug's relative motion when closing on an object.

A searchlight should be mounted over the wheelhouse with the controls in convenient reach of the operator at the control station. A searchlight is handy in shipwork, since the tug operator can illuminate unlighted buoys and obstructions that otherwise would be difficult to see at night. This often is helpful to the pilot as well, since many shiphandlers are reluctant to use the vessel's searchlight, which they feel impairs their night vision. An after control station also should have a searchlight close at hand.

Careful consideration must be given to the use of masthead floodlights, deck lights, and other exterior lighting fixtures on tugs engaged in shipwork. While it may benefit the tug's operator to light up the deck of the tug, misaligned floodlights or searchlights can shine directly into the ship pilot's eyes, obscuring his vision. The same holds true for the ship's lights as they shine on the tug. It can be quite frustrating for the operator of a tug on the towline to look aft to check on his towline angle or the location of the ship's bow and be met by a wall of white light. There are times when the most prudent choice of illumination is to turn off the light. After all, its purpose is to illuminate, not blind.

STRUCTURAL FAULTS ON SHIPS

Two types of general structural problems are associated with ships. One involves design features that interfere with the optimal placement and use of tugs to assist the ship. The second has to do with the ship's having insufficient structural strength to withstand the force of the tugs pushing on the hull or pulling on its bitts or cleats. We discuss both types below

Detrimental Ship Design Features

A number of ship designs have inherent structural features that make them dangerous or difficult for assist tugs. In most cases it is the nature of the ship's service that requires structural peculiarities that make them unhandy for tugs.

Submarines, for example, provide no shipside for the tug's above-water fenders to rest against. Unless the tug has underwater fendering by the bow and alongside the chine, the tug will be hull-to-hull with the sub when it is lying alongside. Aircraft carriers have overhanging superstructures and often many projections along the side that can be hazardous to tugs. Such a vessel usually has few places where a tug can work safely alongside. Some cruise ships and automobile carriers, due to their extended deck houses, have chocks and bitts located fore and aft on abbreviated decks. The flare of the bow and the overhang aft may make it tough for the tug to work there. Some vessels with high freeboard or limited chock locations have thoughtfully provided "pocket chocks" in the ship's side to which a tug can make fast. They are a welcome improvement if they are the right height above the water and located in an area where a tug can work safely. (See fig. 5-14.)

Some of the new vessels with transom sterns are very wide aft to accommodate the large deckhouse. These may have most or all of their chocks located on a small stern deck. If there are chocks elsewhere, they usually are not far enough forward of the tremendous after overhang. Some ships are fitted with constant tension winches and use wire rope or spring lay cable for mooring lines. Rolling chocks usually accompany these installations and

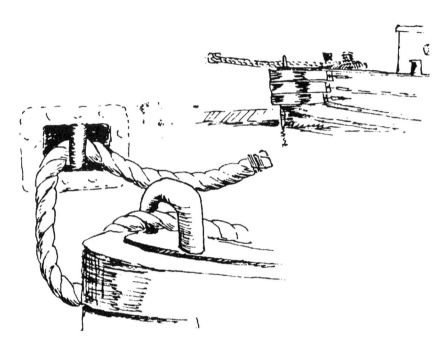

Fig. 5-14. Tug making fast to a recessed or pocket chock in the side of a passenger liner. Note how the line is passed through chock and bitter end is made fast back aboard tug.

often are unsuitable for the tug's working lines, since their apertures are too small. Because bitts are not used for tying up the vessel, none may be available for securing the tug's lines.

A few ship designs locate well-positioned chocks right over the ship's overboard discharges. A tug with a line up to one of these chocks may find main engine cooling water, ballast water, and gray water deposited on its decks. In the days prior to mandatory sewage containment systems, this discharge may have consisted of unmentionable material from the sanitary system. The vertical half-round pipes welded to a ship's side covering waste water and generator discharges also can be a nuisance, as they may fit between the tug's fenders and be damaged. The tire casings used for fendering also can be torn off.

INSUFFICIENT STRUCTURAL STRENGTH

For many years tugs were simply not powerful enough to challenge the ship's structural strength. Ship designers focused their efforts on providing enough structural strength for the ship to function safely in its dedicated service and were less concerned with the potential forces generated by an assist tug.

Today's high-horsepower tugs dedicated to shipwork generate tremendous pulling and pushing forces. At the point of contact, the tug's working end in a push, or the ship's bitts in a pull, these forces can apply stresses to the ship's hull and deck structures that surpass the loads these structures experience in the ship's regular service. Ship's side shells have been pushed in and bitts literally torn off the deck. In response, some ships have incorporated reinforced hull sections at specific locations to accommodate tugs and added oversized bitts integrated into the ship's framework for adequate strength.

Naval architects and marine engineers should consider these factors when designing a ship. If tugs must conform to certain structural requirements to be effective at their trade, it is not unreasonable to expect a vessel requiring their assistance to provide chocks in suitable locations and bitts of sufficient strength so that the tugs can indeed be employed effectively and safely. The safety of the ship itself may depend upon the tug's ability to function efficiently without being exposed to undue risk.

STUDY QUESTIONS

1. Describe the important aspects of the hull of a tug used for shipwork.
2. Is the hull reinforced?
3. Why are deck structures set so far inboard on harbor tugs?
4. What is a bullnose?
5. What is a staple?
6. Where are the bitts located on conventional American harbor tugs?
7. What are the structural requirements for the bitts and deck fittings?
8. What type of mast should be installed on a harbor tug?

9. With respect to visibility, how should the wheelhouse be arranged?
10. Why are model hull single-screw tugs preferred for assisting submarines?
11. What structural characteristics of a ship can make it difficult or dangerous for a tug assisting it?
12. What are pocket or recessed chocks and why are they useful?
13. What are the three functions of fendering?

CHAPTER SIX

GEAR AND RIGGING FOR SHIPWORK

The connecting gear between tug and ship is a critical link in shipwork. Prior to the advent of new line and winch technology, it was considered the weak link. A failure of the connection between tug and ship at the wrong time can have expensive and career-altering consequences. At one time, minor damage to the ship or dock was tolerated in shipwork. Those days are gone. Today even a minor incident can trigger the participation of government agencies, delays in the ship's itinerary, and costly repairs, all at the expense of the shipping company and pilot's reputation. The best interests of tug companies, operators, and ship pilots require that the gear and rigging used in shipwork are up to task.

The traditional approach to building a reliable connection between ship and tug focused on the material, construction, and size of the towline. This somewhat one-dimensional approach has evolved into the towline system approach of today. This is due largely to the tremendous breaking strengths of new synthetic line materials. The breaking strengths of these new, high-tech lines easily can exceed the structural strength of customary tug bitts, winch mechanisms, ship chocks, and cleats. In the past, a towline was matched to fit the performance of the tug. Today it is not uncommon to build the tug to match the performance of the towline.

While still an essential element, the towline is one of several key components to a towline system. The components of a functional and safe system are:

- *Working Lines*
  - o *line material*
  - o *line construction*
- *Fastening/Release Mechanisms*
  - o *bitts, cleats*
  - o *winches*
  - o *tow hooks*
- *Best Practices*
  - o *operational procedures*
  - o *handling and care*

The objective of a good towline system is to make certain the physical connection between tug and ship maximizes the combined horsepower in the wheelhouse and the engine room.

WORKING LINES

A typical complement of lines found aboard tugs in shipwork is:

1. A good headline, about 150–200 feet in length with an eye spliced at each end or with one end attached to a headline winch.
2. Sometimes another slightly smaller line with an eye at one end and the other end whipped, so it can be rove through the chocks that are sometimes installed in the ship's side in passenger vessels and in some automobile carriers that have short forward and afterdecks (fig. 5-14).
3. A good secondary line similar to the headline that can be used as a springline.
4. A towline similar in strength to the headline and springline but of longer length (200 to 250 feet). For ease of handling and to keep it out of the tug's propellers, it is sometimes desirable that this line be of the floating type.
5. A stern line or quarter line (terms used interchangeably). This line can be smaller than the towline because its

purpose is simply to hold the tug in position. This can be for either breasted towing or push/pull towing when backing against the headline. This line might also be of the floating type and, like the towline, must be longer than either the headline or springline.

Lines used in shipwork must meet fluctuating strength, flexibility, durability, and dynamic loading criteria. One type of line material and construction cannot meet all these criteria. Each type of line has pluses and minuses. The technical information surrounding today's ropes may make sense to a rope engineer but may be indecipherable to the user on the deck or pilothouse of the tug. To use them properly and within their designed limits, the tug operator and his deck crew should be versed in the basic terminology and primary characteristics of each type of rope's material and construction.

Types of Working Lines

Towlines are constructed from natural or man-made fibers that are spun into yarns and then twisted into strands to create the body of the rope. In the early years of shipwork manila, derived from natural sources, was the preferred material for the tug's working lines. However, with the increasing size and power of tugs, manila became overmatched. Manila has low strength, becomes very heavy, sinks when wet, and is subject to mildew and dry rot. Once the main material for tug lines in shipwork, manila has been eliminated or relegated to the bosun's locker for heaving lines, messengers, and stoppers. It has been replaced by man-made synthetics that are stronger, more durable, and have a better strength-to-weight ratio.

Not all synthetic lines are suitable for shipwork. Some are recommended only for limited applications where they are not subjected to chafe or exceptionally heavy strains. While synthetics are not vulnerable to mildew, as are natural fibers, they are susceptible to deterioration from prolonged exposure to sunlight or to petroleum products like gasoline etc., and to damage by rust.

Nylon is the name used for a number of polyamide synthetic fibers, and has advantages and disadvantages. It is preferred in applications that require a combination of strength and elasticity. Nylon is the most elastic cordage, and one of the strongest used for commercial marine purposes. It has a good strength-to-weight ratio, resists rot and mildew, and has high elongation properties under strain (30 to 40 percent). However, it sinks when wet and loses approximately 15 to 20 percent of its strength. It can be damaged when exposed to some chemicals and by prolonged exposure to the ultraviolet (UV) rays of sunlight.

Certainly, nylon's strength and shock-absorbing character are well-suited for towing hawsers, shock lines, and emergency towlines. However, its extreme elasticity makes it impractical and dangerous to use as a working line in shipwork. While some give may be desirable in the tug's working lines, the extreme stretch of a pure nylon line renders it impractical for the precision required in shipwork. In addition, the recoil of a parted nylon line under strain can have fatal consequences.

Polyester is almost as strong as nylon, less elastic, and excellent for general-purpose marine applications. Dacron is the trade name for polyester fiber. It has a good strength-to-weight ratio, moderate elongation (15 to 20 percent), excellent abrasion resistance, a high melting point and a low coefficient of friction. This allows it to slide easily around bitts and reduces the chance of the fibers fusing under strain. These characteristics make it suitable for slings, mooring lines, and working lines on tugs.

Polypropylene is a light-weight, low-cost, general-purpose fiber. It floats but is weaker than nylon or polyester, chafes and abrades easily, and has a low melting point. High friction points, such as bitts, may cause the fibers to melt and fuse when the line is surged under heavy strain. Polypropylene fibers are susceptible to damage from UV rays and may disintegrate with prolonged exposure to sunlight. Polypropylene is suitable in light-work applications. It is used extensively as one of the components in material blends.

Blends are combinations of nylon, polyester, and polypropylene fibers marketed under a variety of trade names. Blends use two complementary types of fibers that compensate for each fiber's respective deficiencies. Usually a tough outer shell of nylon or Dacron surrounds an inner core of polypropylene or polyethylene. The end product has greater tensile strength than either material alone, and resists chafe and abrasion. Since the inner core is made from lighter material, it has the desirable quality of floating when wet. Blends result in ropes that have moderate strength and elongation. There are a variety of blended-line products. Each has its own specific characteristics, but most have qualities well-suited for shipwork.

The new synthetics (Aramid and HMPE fibers) have the highest strength-to-weight ratio of any fiber. The breaking strengths of lines made out of the new synthetics far surpass the breaking strength of similar sized lines made of wire or traditional synthetics. The evolution of today's high-horsepower tractor and ASD tug would not have been possible without the introduction of these synthetic materials. Prior to their advent, no practical line technology met the strength, durability, weight, and stretch criteria suitable for an omni-directional, high-horsepower tug in ship escort or assist work.

Although these new synthetics are technically brilliant, the nomenclature surrounding them can be quite confusing. Fiber name, trade names, processing names, and manufacturer product names are used interchangeably.

There are two types of new synthetic fibers, Aramid and high modulus polyethylene (HMPE), each of which is becoming increasingly popular. Aramid fibers are known by their trade names—Kevlar®, from the DuPont Co., and Twaron®, from the Akzo Nobel Co. The trade names for HMPE fibers are Spectra®, from Honeywell (ex Allied-Signal) Co., and Dyneema®, from the DSM Co. Rope manufacturers have developed numerous proprietary fabrication processes using Aramid and HPME fibers. These are marketed under a variety of brand and product names.

However, the interests of the tug operator and shiphandler lie more in what these lines can do rather than what they are called. They are much stronger than the conventional nylon, Dacron, polypropylene, or blends commonly used in the towing industry. Lines made from the HMPE fibers float. They have the same strength as a cable (wire rope) of the same diameter and about one-tenth the weight. They are soft to handle and very flexible. However they are expensive, have a low resistance to heat, and a low friction coefficient.

Table 6-1 lists some of the different general characteristics between ropes derived from Aramid and HMPE fibers.

Characteristic	Fiber Type	
	ARAMID	HMPE
Floats	No	Yes
Strength-to-weight ratio	Excellent	Highest
Abrasion resistance	Fair	Highest
Fiction coefficient	Low	Lowest
Melting point	High	Low
UV resistance	Fair	Excellent
Loss of strength when wet	5%	0%

Table 6-1. Comparative characteristics of new synthetics.

Coatings are another variable in the manufacturing of all lines, regardless of material source. Different types of coatings are applied to the individual fibers, yarns and ropes during the production process. Their purpose is twofold. One is to protect the rope from external sources of damage, such as chemicals, UV rays, and prolonged exposure to water. The other is to reduce the internal friction between rope strands as they twist and elongate under strain. Internal abrasion is a significant factor in weakening a rope and shortening its useful life.

Table 6-2. Table of fiber rope strengths for working lines and hawser. *(Courtesy of Wall Rope Company.)*

Size		*Manila*		*Polypropylene (Monofilament)*		*POLY-plus*		*POLY-cron*		*Nylon*		*Dacron (Polyester)*	
Dia.	*Cir.*	*Tensile Strength*	*Lbs. per 100 ft.*	*Tensile Strength*	*Lbs. per 100 ft.*	*Tensile Strength*	*Lbs. per 100 ft.*	*Tensile Strength*	*Lbs. per 100 ft.*	*Tensile Strength*	*Lbs. per 100 ft.*	*Tensile Strength*	*Lbs. per 100 ft.*
3/16"	5/8"	405	1.5	800	.7	—	—	—	—	1,000	1.0	1,000	1.2
1/4"	3/4"	540	2.0	1,250	1.2	—	—	—	—	1,650	1.5	1,650	2.0
5/16"	1"	900	2.9	1,900	1.8	—	—	—	—	2,550	2.5	2,550	3.1
3/8"	1 1/8"	1,215	4.1	2,700	2.8	2,650	3.5	—	—	3,700	3.5	3,700	4.5
7/16"	1 1/4"	1,575	5.25	3,500	3.8	3,600	5.0	—	—	5,000	5.0	5,000	6.2
1/2"	1 1/2"	2,385	7.5	4,200	4.7	4,500	6.5	—	—	6,400	6.5	6,400	8.0
9/16"	1 3/4"	3,105	10.4	5,100	6.1	5,450	7.9	—	—	8,000	8.3	8,000	10.2
5/8"	2"	3,960	13.3	6,200	7.5	6,400	9.4	—	—	10,400	10.5	10,000	13.0
3/4"	2 1/4"	4,860	16.7	8,500	10.7	8,400	12.0	—	—	14,200	14.5	12,500	17.5
11/16"	2 1/2"	5,850	19.5	9,900	12.7	10,200	14.5	—	—	17,000	17.0	15,500	21.0
7/8"	2 3/4"	6,930	22.5	11,500	15.0	12,000	17.0	—	—	20,000	20.0	18,000	25.0
1"	3"	8,100	27.0	14,000	18.0	15,000	21.5	14,000	26.5	25,000	26.0	22,000	30.5
1 1/16"	3 1/4"	9,450	31.3	16,000	20.4	17,100	24.2	—	—	28,800	29.0	25,500	34.5
1 1/8"	3 1/2"	10,800	36.0	18,300	23.7	19,300	27.0	21,000	34.0	33,000	34.0	29,500	40.0
1 1/4"	3 3/4"	12,150	41.8	21,000	27.0	22,000	32.5	24,000	39.0	37,500	40.0	33,200	46.3
1 3/16"	4"	13,500	48.0	23,500	30.5	25,000	38.0	27,000	44.0	43,000	45.0	37,500	52.5
1 1/2"	4 1/2"	16,650	60.0	29,700	38.5	31,300	46.0	34,000	55.0	53,000	55.0	46,800	66.8
1 5/8"	5"	20,250	74.4	36,000	47.5	38,300	55.0	42,000	67.0	65,000	68.0	57,000	82.0
1 3/4"	5 1/2"	23,850	89.5	43,000	57.0	46,500	65.0	50,000	80.0	78,000	83.0	67,800	98.0
2"	6"	27,900	108.0	52,000	69.0	56,500	83.0	60,000	95.0	92,000	95.0	80,000	118.0
2 1/8"	6 1/2"	32,400	125.0	61,000	80.0	65,500	97.0	70,000	112.0	106,000	109.0	92,000	135.0
2 1/4"	7"	36,900	146.0	69,000	92.0	74,000	108.0	80,000	127.0	125,000	129.0	107,000	157.0
2 1/2"	7 1/2"	41,850	167.0	80,000	107.0	86,000	122.0	92,000	147.0	140,000	149.0	122,000	181.0
2 5/8"	8"	46,800	191.0	90,000	120.0	96,000	138.0	105,000	165.0	162,000	168.0	137,000	205.0
2 7/8"	8 1/2"	52,200	215.0	101,000	137.0	105,000	155.0	—	—	180,000	189.0	154,000	230.0
3"	9"	57,600	242.0	114,000	153.0	122,000	179.0	130,000	208.0	200,000	210.0	174,000	258.0
3 1/4"	10"	69,300	299.0	137,000	190.0	144,000	210.0	160,000	253.0	250,000	263.0	210,000	318.0
3 1/2"	11"	81,900	367.0	162,000	232.0	170,000	248.0	—	—	300,000	316.0	154,000	384.0
4"	12"	94,500	436.0	190,000	275.0	200,000	290.0	—	—	360,000	379.0	300,000	460.0

Nominal Diameter		*Size Number (Circ)*	*Approximate Weight*		*Minimum Tensile Strength*	
Inch	*MM*		*Lbs/100 ft*	*Kg/100m*	*Pounds*	*Kilograms*
			PLASMA 12 STRAND			
7/16	11	1¼	4.2	6.3	21,000	9,500
½	12	1½	6.4	9.5	31,300	14,200
9/16	14	1¾	7.9	11.8	37,900	17,200
⅝	16	2	10.6	15.8	51,400	23,300
¾	18	2¼	13.3	19.8	68,500	31,100
⅞	22	2¾	19.6	28.8	92,600	42,000
1	24	3	23.4	34.9	110,000	49,900
1⅛	28	3½	31.9	47.6	147,000	66,700
1¼	30	3¾	36.2	54.0	165,000	74,800
1 5/16	32	4	41.7	62.2	196,000	88,900
1½	36	4½	51.7	77.1	221,000	100,000
1⅝	40	5	65.7	98.0	291,000	132,000
1¾	44	5½	78.4	117	314,000	142,000
2	48	6	91.4	136	355,000	161,000
2⅛	52	6½	109	163	428,000	194,000
2¼	56	7	122	182	481,000	218,000
2⅝	64	8	167	249	560,000	254,000
3	72	9	214	307	780,000	354,000
3¼	80	10	261	389	940,000	427,000
3⅝	88	11	324	483	1,250,000	567,000
4	96	12	394	587	1,520,000	690,000
			SPECTRA 12 STRAND			
⅜	9	1⅛	3.7	5.5	13,900	6,300
7/16	11	1¼	4.2	6.5	14,800	6,700
½	12	1½	6.4	9.5	22,500	10,200
9/16	14	1¾	7.9	11.8	27,700	12,600
⅝	16	2	10.6	15.8	36,600	16,600
¾	18	2¼	13.3	19.8	43,200	19,600
⅞	22	2¾	19.6	28.8	61,000	27,600
1	24	3	23.4	34.9	72,000	32,600
1⅛	28	3½	31.9	47.6	91,800	41,600
1¼	30	3¾	36.2	54.0	102,600	46,500
1 5/16	32	4	41.7	62.2	114,300	51,800
1½	36	4½	51.7	77.1	141,300	64,000
1⅝	40	5	65.7	98.0	167,400	75,900
1¾	44	5½	78.4	117	198,000	89,700
2	48	6	91.4	136	225,000	102,000
2⅛	52	6½	109	163	270,000	122,000
2¼	56	7	122	182	317,700	144,000
2½	60	7½	148	221	360,000	163,000
2⅝	64	8	167	249	370,800	168,000
2¾	68	8½	187	279	405,000	184,000
3	72	9	214	307	508,500	230,000
3¼	80	10	261	389	616,500	279,000
3⅝	88	11	324	483	765,000	347,000
4	96	12	394	587	900,000	408,000

Tensile strengths are determined in accordance with ASTM test method D 4268-83. Weights are calculated at linear density under standard preload ($200d^2$) plus 4%.

Table 6-3. Comparative strengths of spectra and plasma twelve-strand synthetic fiber lines. *(Courtesy of Puget Sound Rope.)*

LINE CONSTRUCTION

The pattern and lay up of the fibers is as important to the strength and durability characteristics of a rope as the source of the fiber's material. There are five common rope construction patterns:

- three-strand
- six-strand
- eight-strand
- twelve-strand
- double-braid

Three-strand is the most common type of twisted rope. It has good abrasion resistance but has a tendency to hockle or kink, which weakens the line. Because the rope strands twist in the same direction (lay) (fig. 6-1) they are prone to kinks if coiled in the wrong direction.

Six-strand also is a twisted rope but has less tendency to hockle than a three-strand rope.

Eight-strand ropes are made up of four pairs of two strands plaited (braided) in a square pattern (fig. 6-1). This pattern creates a rope that is more durable than three-strand, is very flexible and immune from kinks. It has high shock-absorbing qualities.

Twelve-strand rope is a braided rope consisting of twelve twisted strands braided into a single primary rope braid. This construction has the benefit of being easily spliced and repaired and handles compression loads particularly well. Compression loads occur when the rope is led around staples, pulleys, winches, or other rounded objects. Twelve-strand construction commonly is combined with HMPE fibers to produce the high strength towlines used on today's tractor and ASD ship escort tugs.

Double-braid rope consists of an inner core rope and an outer cover or jacket rope (fig. 6-1). Both ropes are braided and contribute to the strength of the rope. The materials of each contributory rope are picked to meet the strength, stretch, durability, and floatability criteria of the finished rope.

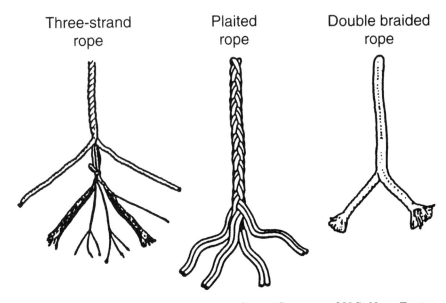

Fig. 6-1. Standard synthetic rope constructions. *(Courtesy of U.S. Navy Towing Manual.)*

WIRE ROPE

Wire rope is commonly used in shipwork for pendants and working lines. Tugs in some areas make up alongside ships with wire rope. Wire rope also is used to construct pendants, fifteen to twenty-five feet in length, that are attached to the outboard end of fiber working lines. This protects the soft line from the chafe and abrasion that occur where the lines pass through the ship's chocks. Pendants normally are attached to only the headline or springline since the weight of the pendant would cause a towline or quarter line to sink, making them difficult to retrieve.

Wire rope pendants and working lines should have eyes large enough to fit over a ship's bitts and can be spliced or swaged on the outboard ends. The pendant also has a small eye in the other end that attaches to the soft line.

Tow Cable Strengths

6 × 19 Class Wire Rope

Breaking Strength, Tons of 2,000 lbs.

Rope Diameter in inches	Xtra Improved Plow Steel		Improved Plow Steel	
	Fibre Core	IWRC*	Fibre Core	IWRC
½"	—	13.3	10.7	11.5
⅝"	—	20.6	16.7	17.9
¾"	—	29.4	23.8	25.6
⅞"	—	39.8	32.2	34.6
1"	—	51.7	41.8	44.9
1¼"	71.0	79.9	64.6	69.4
1½"	101.0	114.0	92.0	98.9

6 × 37 Class Wire Rope

Breaking Strength, Tons of 2,000 lbs.

Rope Diameter in inches	Xtra Improved Plow Steel		Improved Plow Steel	
	Fibre Core	IWRC	Fibre Core	IWRC
½"	—	13.3	10.7	11.5
⅝"	—	20.6	16.7	17.9
¾"	—	29.4	23.8	25.6
⅞"	—	39.8	32.2	34.6
1"	—	51.7	41.8	44.9
1¼"	67.7	76.1	61.5	66.1
1½"	96.6	108.0	87.9	94.5

Source: Bethlehem Steel Co., Wire Rope Division
* IWRC (internal wire rope core)

Table 6-4. Lists the strength of wire rope typically used in shipwork.

Material and Construction

Wire rope for shipwork should be constructed from Improved Plow Steel or Extra Improved Plow Steel for strength. Wire rope construction consists of small wires twisted to form strands that are wound around a central core. The core can be either rope or steel. The terminology used to describe wire rope construction refers to the number of strands in the rope, the number of wires that make up a strand, and the type of core. A 6 x 19 independent wire rope core

(IWRC) is a wire made up of six strands, each strand composed of nineteen wires, layed up around an IWRC.

The more wires used to form a strand the greater the rope's flexibility and resistance to fatigue. However, more wires reduce abrasion resistance. Wire rope used for working lines and pendants should be of 6 x 19 or 6 x 37 construction (6 x 37 is preferable since it is more flexible). A fiber as opposed to steel core is preferred, as the extra flexibility of fiber facilitates handling these small wires by hand. In addition, the fiber core in six-strand cable provides lubrication and some elasticity to the finished wire rope

When wire rope is used, it should be comparable in strength to that of the soft line used.

COMPOSITION WORKING LINES

A composition working line is essentially a line system built out of components of different line materials and construction methods. Composition lines incorporate the same principle used in creating fiber blends, each component's strong features compensating for another's respective weaknesses.

In general, the purpose of composition lines is to:

- use abrasion resistant components at high friction points (ship's bitt and chock)
- create an easily replaced, disposable component in areas of high wear
- accommodate dynamic loading
- create an inherent and purposeful weak link in the line
- provide flexibility and softness to working ends handled manually

These purposes are met by marrying:

- fiber rope and wire pendants

- fiber rope and fiber rope pendants
- wire rope with fiber pendants

They can be designed to meet the demands of a specific type of service (e.g., ship escort), class of tug (e.g., tractor), vessel, or even the experience level of an individual operator.

As an example, a line made of Spectra® or Dyneema® has very low stretch. The elongation properties of an added short polyester or nylon pendant add some accommodation for dynamic shock loads and provide a replaceable component at the end of the line. In addition, the chosen pendant diameter can function as a purposeful weak link, designed to release the tug at load beneath the high breaking strength of the Spectra® or Dyneema®.

Many rope manufacturers and individual tug companies have created unique composition lines with innovative, proprietary splicing techniques that marry different components. It is in creating composition lines that the science of rope making becomes an art. The right combination of line material and design for a composition line requires the creative input of rope manufacturers, tug companies, and individual tug operators. It is this collaboration between the science of rope making, the collective experience of the tug company and the individual hands-on experience of the tug operator that creates a functioning composition line.

LINE FASTENING AND RELEASE MECHANISMS

The connection between ship and tug must be reliable and controllable. Line fastening and release mechanisms must give the tug operator discretionary ability to hold, surge or release the line. While an unexpected broken connection between tug and ship may have expensive consequences for the ship and its pilot, the inability to surge or release the connection may have serious safety consequences for the tug and its crew. The connection between ship and tug has two terminal ends, one on the ship and one on the tug.

THE SHIP'S END OF THE TOWLINE

The fastening mechanism at the ship end of the towline is single purpose. It is devoted strictly to a reliable securing of the tug's line. This requires two components: the bitt or cleat that holds the line and the chock that fairleads the line.

Ship chocks should be well faired, durable, and smooth. Chafe is a prime source of damaged and parted lines. Tug operators should be wary of passing fiber lines through ship chocks that are used primarily for wire rope. The wire may have grooved the bearing surface of the chock creating a rough and abrasive surface. This is an instance that would call for a composition line with a wire pendant.

Both bitts and chocks used on the ship should be of sufficient strength to safely withstand the combined force of the tug's bollard pull, weight, and hydrodynamic resistance. The increasing popularity of the new HMPE synthetic lines has brought to light many ship's deficiencies in this regard. While much of the focus has been on the structural strength of bitts, the chocks are equally important. They are subject to tremendous sheering loads when they serve as the fairlead for a towline attached to a high-horsepower tug engaged in ship escort work.

THE TUG'S END OF THE TOWLINE

At the tug's end of the towline, the fastening mechanism must fulfill at least two and preferably four functions. At a minimum, it must first provide a means to secure the line and second, a means to quickly release it from the tug. Additionally it would be desirable to provide the ability to surge the line under strain, and adjust the length of the towline.

The traditional means of fulfilling these requirements was to have a deckhand make off the line by taking wraps on the tug's bitt or cleat. Deck personnel pulled in or paid out slack lines and released or surged the line by taking wraps off the bitt. However, this is an unsafe, antiquated and increasingly impractical means of securing lines to tugs. First, the deckhand will be working in the danger zone around lines under tremendous strain. Second, many of the new, high-tech ropes slip their wraps off bitts when under strain. Third, there are simply faster and much more reliable mechanical means of accomplishing these tasks.

Tow Hooks

Tow hooks are used widely in Europe as means of securing and releasing towlines. The eye of the tug's towline or a ship's hawser is secured in the hook and locked in place by hydraulic or mechanical means. Tow hooks are set up with a quick-release mechanism that can be operated locally at the hook or remotely from the wheelhouse.

Tow Straps

In the U.S., a quick release or tow strap is sometimes used as a substitute for a towing hook. The strap consists of a short line (six to eight fathoms) with an eye at one end that is rove double through the eye of a ship's hawser and secured to the after tow bitts (fig. 6-2). This permits the crew to cast off the hawser in an emergency.

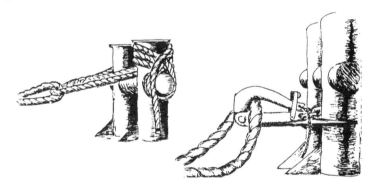

Fig. 6-2. Showing tow hook and an alternate method of using a quick-release strap.

WINCHES

The winch is the best mechanical means for giving line control to the tug operator. Winches may be driven mechanically, hydraulically, or electrically. They may be single, split, or double drums arranged in a configuration that meets the specific demands of different towing services. Regardless of design, the following are the main characteristics required of winches used in shipwork:

- braking capacity
- maximum line-pull capacity
- slack line speed
- remote release

Braking Capacity

Loads on tugs' working lines in shipwork can easily exceed the bollard pull of the tug. A winch brake is one component of a system for managing those loads. The braking capacity must be relative to the type and size of towline used. The winch brake must complement the line's breaking strength and built-in safety factor. Some operators prefer a winch with braking capacity lower than the breaking strength of the line. This ensures that the brake will slip before the line parts, extending the life of the line. Other operators prefer that the brake exceed the breaking strength of the line to make certain that the brake won't slip at a critical moment in a ship assist or escort maneuver.

Some braking systems have an adjustable braking capacity that can be set by the operator. This feature is particularly useful in ship escort work to help manage the dynamic loading of the towline.

Maximum Pull Capacity

There are circumstances in shipwork when it is desirable to shorten the towline while under load. Tugs assisting ships from open water to narrow channels or berths commonly encounter wind or current conditions where there is too little time or room to permit the tug to slow down and shorten up a slack line without the ship drifting into jeopardy. A maximum pull rating of one-half to the entire bollard pull of the tug is common on winches used in shipwork.

Slack Line Speed

Tugs in shipwork commonly use headlines or towlines that are 100 to 200 feet in length. The winch's capacity to rapidly pull in a slack line adds a necessary safety and efficiency feature to tugs in shipwork. A good slack line speed allows the tug to rapidly retrieve a line cast off. This reduces the risk of the line fouling the propeller of the ship or tug. Fast slack line speeds also facilitate the efficiency of tractor or ASD tugs transitioning from a towing to a push/pull position.

Remote Release

Good winch design incorporates the safety feature of a remote quick-release mechanism. From his station in the wheelhouse, the tug operator, should be able to release the winch brake or towline quickly in the event the tug becomes girt, tripped, or in irons.

The line handling winches built for shipwork today are high-performance machines. The popularity of the new synthetic lines in shipwork spurred an evolution in winch technology. The high cost and breaking strength of the new synthetics combined with the development of ship escort maneuvers required a winch that would fulfill multiple functions:

- minimize damage to the line
- pay in pay out line under load
- automatically render and recover at preset load levels
- have high slack line speeds

Winch technology continues to evolve rapidly. Water-cooled brakes, electric winch drives, under-deck installation, tensionometers, and computer-assisted controls are among the recent advances. However, winch performance is not solely a function of technology. Sophisticated machinery requires commensurate training of the winch operator. The winch may be capable of high performance, but if it is not matched by the operator's skill, it can be no more than an expensive powered storage reel with a brake.

BEST PRACTICES

The longevity and performance of the tug's connecting gear depend on more than well-designed and high-performance equipment. They also require best practices to ensure that the gear is used within its designed parameters and maintained appropriately. The risk to personnel and equipment and the cost of replacement warrants vigilant handling and care of the tug's connecting equipment. Today's towline constructed of HMPE fiber may cost upwards of $50,000. At that price, it is imperative that connecting gear be maintained and handled properly.

OPERATIONAL PRACTICES

Choose the right rope for the right job:

- The rope should be of a material, design, and size to ensure safety in the chosen application.
- In calculating a safety factor, consider the anticipated working load, dynamic loads, abrasive surfaces, and environmental conditions.
- Safety factors may vary from 3:1 to as much as 20:1, depending on the level of risk to equipment and personnel.
- Safety factors for a specific application should be determined by professionals who have extensive hands-on experience in the application for which the rope is being designed, and comprehensive knowledge of the rope material and construction performance properties.
- Take into account the predicted reduction in residual towline strength over time. One study found an average reduction of almost 40 percent in towline strength after approximately 1,200 ship jobs (two years)[1].

---

[1] *Residual Strength Testing of Dyneema Fiber Tuglines*. Phil Roberts, Danielle Stenvers, Paul Smeets, Martin Vlasboom. Paper International Tug and Salvage Convention 2002.

Use the rope properly:

- Use maneuvering finesse to increase or ease the load on a line.
- Avoid shock-loading the line through rough or careless tug handling.
- Avoid bending the rope around small radius corners or pulleys.
- Protect the rope at contact points from chafe and abrasion.
- Use the proper splicing technique rather than a knot to connect or terminate the rope.
- Make sure that bitts and winch drums are of sufficient diameter (six to eight times the rope diameter) to minimize wear and fatigue caused by rope bending.
- If possible, avoid steep vertical towline angles (greater than 40 degrees). (Towline forces on steep towlines may be more than three times the tug's bollard pull.)

HANDLING AND CARE

Inspect the rope regularly for:

- kinks or twists
- worn or abraded areas
- cut or frayed strands (inner and outer)
- melted or fused strands (inner and outer)
- damage from chemical exposure
- splice movement
- glossy or glazed areas
- inconsistent diameter (flat spots or bumps)

Keep a line log:

- Note number and duration of jobs
- Maintain a load history
- Note occurrences of shock loads

- Record frequency and results of inspections
- Note instances of major maintenance (end for end, new splices)

Stow line properly:

- in a clean, dry, well-ventilated area
- away from heat sources
- away from paint or thinner vapors
- away from sunlight
- coiled properly

Know when to retire or downgrade a line:

- There are no definitive rules or industry standards.
- Develop an application specific set of criteria for retiring or downgrading a line based on experience with the line's application, rope's specific load history and inspection results.

The dependability of the connection between tug and ship rides on more than the towline. Tug operator and ship pilot must have confidence that the tug's towline system's equipment is adequate for the job at hand, and that the tug operator is trained in the proper use and maintenance of this equipment. Only when both pilot and tug operator trust the connection between tug and ship can they use their expertise to the fullest capacity.

AUXILIARY EQUIPMENT

The following items of equipment may seem inconsequential, but without them shipwork can be more difficult or dangerous.

Heaving lines are used to pass a tug's working lines to a ship. On tugs they are usually about fifteen fathoms long. If they are tied directly to the tug's working line they should be at least three-eighths inch to seven-sixteenth inch in diameter, and made of manila or some other material that is not slippery. This provides a decent handhold for the seamen on deck of the ship. If tag lines are used, heaving lines often are made of braided line a quarter-inch to five-sixteenth-inch in diameter. These can be thrown farther and higher than thicker lines, which can be helpful on windy days when working a ship that is light with a lot of freeboard.

Tag lines are lengths of line secured to the outboard end of the tug's working lines and used to provide a handhold for the ship's deck force to pull them on board the vessel. Tag lines are attached to the bitter end of the heaving line, which is passed to the vessel first. They are especially helpful in hauling up large working lines and lines with an attached wire rope pendant. Tag lines are typically made from three-quarter inch to one inch diameter non-slippery rope and are thirty to forty feet long. The length is to enable the ship's crew to pull in the heaving line and get a hand on the tag line before the full weight of the hawser is felt. Larger ships with more freeboard require longer tag line lengths.

Goblines are straps that pass over or are shackled to a towline. They act as a preventer to keep the towline from leading over the side and tripping or capsizing the tug. Goblines are secured to the afterdeck near the stern of the tug and permit the towline to swing only through a limited arc (fig. 6-3). Goblines are not often seen on harbor tugs in the U.S., but are frequently used on tugs doing shipwork in Europe. They often are employed with tow hooks and use winches to adjust the gobline's length. This allows the tug operator to mechanically move the tug's towing point. When used on offshore tugs in this country, they are sometimes called tie downs or hold downs.

Some types of additional gear can make tugs handier, more efficient, and safer. They include gratings to keep the working lines dry and off of the deck; tarpaulins to protect winches and lines against spray, sunlight and freezing spray; nonskid decks that make for safer footing; and a few bags of salt to melt the ice in winter. Some large tugs have powered drums to handle and stow their working lines. These are a welcome addition when handling fourteen-inch or even ten-inch lines, especially when they are wet.

These kinds of amenities are the mark of a tug designed with the eye of a professional mariner.

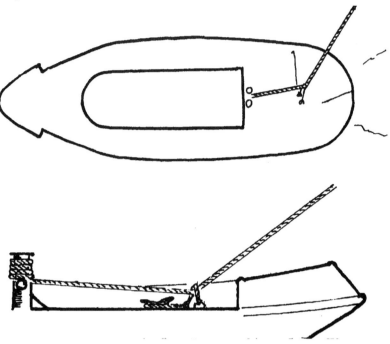

Fig. 6-3: A gobline rigged to restrict the movement of the towline.

STUDY QUESTIONS

1. What are the components of a towline system?
2. What types of synthetic lines are used in shipwork?
3. Are 100 percent nylon lines used for shipwork?
4. What are Dacron lines used for primarily?
5. What are polypropylene and polyethylene lines used for?
6. What are blended lines used for?
7. What are the two common trade names for HMPE fibers?
8. What purpose do wire rope pendants serve?
9. What are the five common rope construction patterns?
10. What types of wire rope are suitable for pendants and working lines?
11. What are the purposes of a composition line?

12. What are the four important characteristics of a winch?
13. What are the two main operational Best Practices?
14. List the common items of auxiliary equipment.
15. Describe the heaving lines commonly used on tugs.
16. Describe and explain the purpose of a tag line.
17. Can tag lines be made of manila?
18. What is a gobline and what is its purpose?
19. What is a quick-release strap, and how is it used?

CHAPTER SEVEN

HANDLING THE LIGHT TUG

Learning to handle a tug in shipwork requires competence at multiple levels of skill. The first and simplest level is handling the tug by itself in a variety of situations. Once proficient at this level, increasingly complex maneuvering tasks can be added to the operator's skills.

This chapter describes the fundamental principles of handling light tugs. Learning to handle a light tug always includes hands-on practice. But practice without an understanding of principle is an incomplete learning experience. Knowing what principle caused the tug to move and *how* to make the tug move are equally important in any maneuver.

Tug handling and shiphandling share many similarities. Both rely on sensing a vessel's motion, mentally envisioning where this motion will lead, deciding what changes in motion may be required, identifying the appropriate maneuvering technique, and using the vessel's steering and propulsion to execute the selected technique. Both rely on a combination of intuition and thinking ahead.

In the early phase of learning an operator consciously goes through this process, predominately thinking through a maneuver. The operator systematically assesses the vessel's motion, makes an educated guess where the vessel will end up, weighs the different maneuvering options, mentally runs through how to execute those options, tentatively moves the engine and steering controls, and observes if the decisions were correct. Not uncommonly, the operator may find that taking too long to manipulate the controls has allowed the tug to pass into a new set of circumstances. This is why anyone learning to handle a tug is encouraged to go slow and allow enough time for conscious processing.

An expert tug handler thinks with his hands, and the processing appears seamless and intuitive. The expert's perception and processing of the vessel's motion is immediate. This is a result of familiarity with the vessel and understanding maneuvering principles. The expert knows the engine response time, rudder power, stopping distance, and other critical characteristics of the tug. An in-depth understanding of tug handling dynamics serves as the foundation for applying that knowledge. Perception and comprehension are connected directly to the hands; assessment and action appear almost simultaneous.

For these reasons, learning to handle a tug is a hands-on process. A new operator can hone intuitive skills only by cultivating a feel for the tug through practice. There is no convenient rule of thumb for estimating a tug's turning rate or stopping distance. It can be acquired only by handling the tug and gaining a feel for how quickly a particular tug turns or stops. Even sister ships may differ in rudder power and engine power, both ahead and astern, and how long they take to respond to the controls.

Seasoned handlers know this and can quickly feel out an unfamiliar tug. Most handlers will try a few maneuvers with a tug before working in close quarters. A new operator should practice handling the tug in an unrestricted area until developing a feel for the tug, a judgment of distance, and the tug's stopping and turning characteristics. To sharpen perceptions, it helps to work close to a reference point (a buoy or the corner of a dock). Touch-and-go landings also are helpful in developing this skill.

Once an operator is proficient at operating the tug in open water and has a feel for the tug's responses, he is ready to undertake the next step: applying open-water skills to specific close-quarters maneuvers. Competent application of maneuvering fundamentals requires knowledge and foresight in four areas.

First, an operator must know *what* he wants the tug to do. This may sound easy and obvious, but in practice can be complex and unclear. It is easy to visualize the end result of a docking maneuver:

the tug is stopped and in position at the berth. But there are many options in getting there. The operator must decide on the speed and angle of approach, judge the tug's momentum through the water, and what it will take to bring it to a stop. In essence, the operator must establish a mental transit plan before undertaking the maneuver.

Second, an operator must know *how* to make the tug follow the transit plan and how to manipulate the tug's controls to make the tug turn, slow down, reverse direction, or stop as required by the transit plan.

Third, the operator must know *when* to execute the control actions that will change the tug's course and speed, and anticipate the required changes in the tug's direction and speed.

Fourth, and most critically, the operator must constantly *adjust* his plan and actions. Tug handling is dynamic. Although many maneuvering situations appear similar, no two are ever alike. A tug may come into its home berth hundreds of times, but each approach and series of maneuvers will be slightly different. The operator must be aware of the nuances of difference and adjust accordingly.

Combining these four elements produces a series of critical decision points (CDPs). These are the points in a maneuvering sequence that require critical action at a specific time. Any maneuver is a series of CDPs; points in which an action missed or ill-timed, exponentially exacerbates the difficulty of the maneuver. A simple docking maneuver illustrates this concept. The vessel handler mentally calculates the optimal angle of approach to the dock. However if the operator misses the helm over point, his preparatory work is nullified. The vessel will carry past the point of execution, requiring a steeper angle of approach, more rudder angle to get the vessel turned, and, in some cases, more speed on the approach. This is where the experienced operator makes appropriate adjustments and the inexperienced operator may flounder (fig. 7-1).

Identifying critical decision points and taking the corrective actions requires a thorough understanding of the handling principles associated with various types of tugs. These principles and basic maneuvers are discussed below.

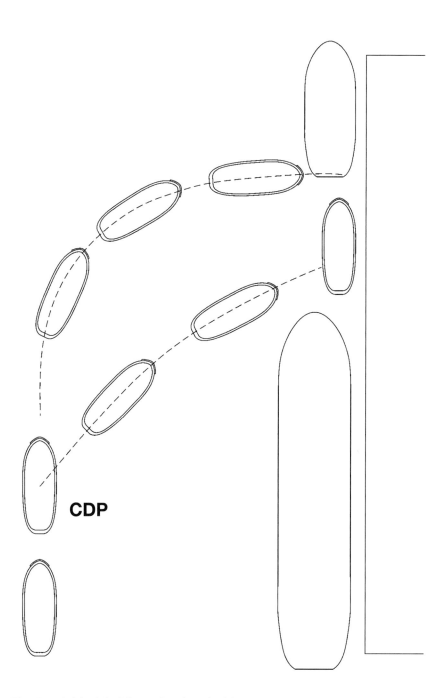

Fig. 7-1. Critical decision points in a docking maneuver.

SINGLE-SCREW, RIGHT-HAND PROPELLER

Most seasoned tug handlers claim that the best tool to learn tug handling is the conventional, single-screw tug. This is because the single-screw tug represents the forerunner of all tug designs. The fundamental principles of maneuvering a tug with one propeller and one engine have universal application. An understanding of how to maneuver a single-screw tug is the foundation for handling the more maneuverable twin-screw, tractor, and ASD designs.

One of the elemental factors in handling a single-screw tug is propeller torque or propeller walk. When a propeller produces thrust it is not solely in a fore or aft direction. In fact, the propeller's tendency is to turn the vessel as well as propel it.

There are many technical descriptions of propeller torque. For the tug operator it is more important to understand what will happen rather than why it will happen. As a propeller rotates it has a tendency to "walk" the stern in one direction. Walking can best be described by the following image (fig. 7-2). From a view point aft, a propeller acts like a paddle wheel when it passes through the bottom of its rotation. The paddle wheel effect will move the stern to one side or the other, depending on a propeller's rotation.

From this viewpoint aft of the tug, a propeller appears to rotate in a clockwise or counterclockwise direction. When a propeller is delivering ahead thrust a clockwise rotation is referred to as a "right-hand propeller," counter clockwise is referred to as a "left hand propeller." The names come from the direction the propeller blades are moving at the top of their rotation.

A right-hand propeller has a tendency to move the stern to starboard and the bow to port when engaged ahead. When reversed, it moves the stern to port and the bow falls off to starboard (fig. 7-3).

The tug operator can easily use the rudder to overcome propeller torque effects when going ahead. This is because a single-screw tug rudder is in the discharge wash of the propeller, its most effective position. When going astern the torque effect predominates since the rudder lies on the intake side of the propeller. Only after the tug has gained sternway and water begins to consistently flow around the rudder can a single-screw tug overcome the effect of propeller torque.

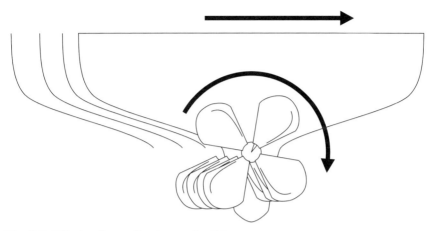

Fig. 7-2. Effects of propeller torque (walk).

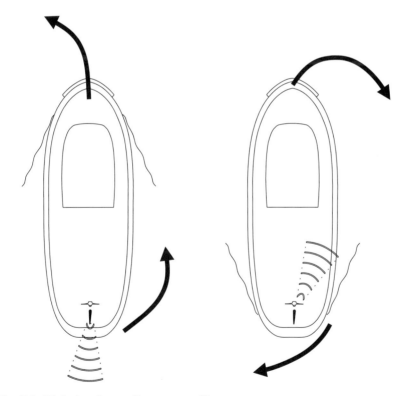

Fig. 7-3. Right-hand propeller torque effects.

MANEUVERING THE SINGLE-SCREW TUG

The following descriptions assume the tug begins dead in the water. When the engine is engaged ahead with the rudder amidships, it moves ahead and starts turning slowly to port as a result of propeller torque. To stop or reverse the direction of movement, the engines are simply engaged in the opposite direction until the desired effect is achieved. If the tug's rudder is turned right or left with the engine engaged ahead, the tug immediately begins to turn in the same direction as that of the helm. The rate of turn depends on the amount of rudder angle used and the engine speed. The tug usually turns in a smaller circle at slow speed, but this depends on the tug.

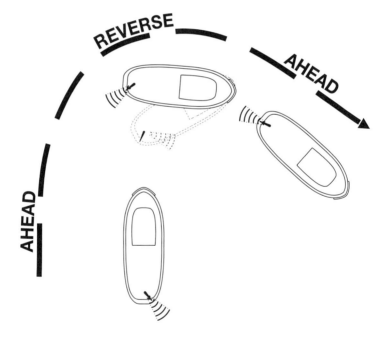

Fig. 7-4. Single-screw tug turning around.

In the same situation, if the engine is engaged astern, the tug moves astern and backs to port because of the propeller's torque. After gaining sufficient sternway for the rudder to take effect, some tugs steer when backing and moving astern. These tugs usually continue to steer when moving astern with the engine stopped until sternway is lost. Many tugs, however, do not respond well when backed. They can be controlled only by turning the rudder in the appropriate direction and "kicking" the engine ahead from time to time. In this instance the engine is used ahead only enough to assist the steering while maintaining sternway.

When a single-screw tug is required to make a hard turn in a confined space, the helm is put over in the required direction, and the engine is engaged ahead. After the tug has started to swing smartly but before it gathers too much headway, the engine is backed just enough to kill the headway without reducing the swing. The helm usually is left untouched since its position is immaterial until the tug has sternway. A tug with a right-hand propeller turns much more readily to starboard since the torque of the propeller assists by kicking the stern to port when the engine is backed (fig. 7-4).

This maneuver, which also is called "back and fill," may have to be repeated several times until the tug has completed the turn. Its success can depend on the number of maneuvers that can be accommodated by the tug's engine and control system. A direct-reversing diesel with a small air capacity may allow only four or five quick starts before requiring time to recharge the air bottles. This may leave the operator in an embarrassing and untenable situation: halfway through a turn in a confined area, drifting down on a dock or other vessels, waiting for the air bottles to recharge.

Undocking a single-screw tug usually involves swinging the stern out and backing clear of the berth (fig. 7-5). The tug will cast off all of its lines, except its forward springline, then come ahead easily into the springline with its helm turned toward the dock. When the stern has opened up enough, the tug casts off the springline and backs clear of the dock.

Docking a single-screw tug must take into account the expected reaction of the tug to propeller torque when backing down. A single-screw tug normally approaches the dock slowly at a 15 to 20° angle, stops its engines when it is close to the dock, and then bears off. It can back its engines when it is close alongside, put out a forward

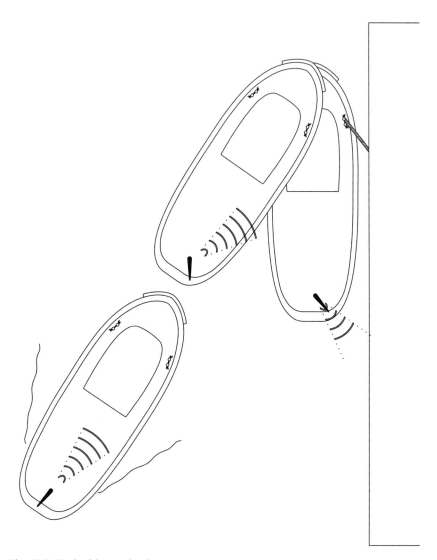

Fig. 7-5. Undocking a single-screw tug.

springline, and work easy ahead on the springline until it is in
position alongside (fig. 7-6).

A seasoned operator of a single-screw tug can always be
identified by the deft use of propeller torque to best advantage.
One example of this is landing a single-screw tug with a right-hand

propeller, starboard side to a dock, without using a springline. An experienced operator will approach the dock at a fairly shallow angle, then put the rudder hard left, give a "kick" ahead to get the bow to open on the dock, and back hard as the stern swings in toward the dock. The propeller torque checks the swing of the stern, closes the bow on the dock, and brings the tug to a stop neatly in position to put lines over (fig. 7-7).

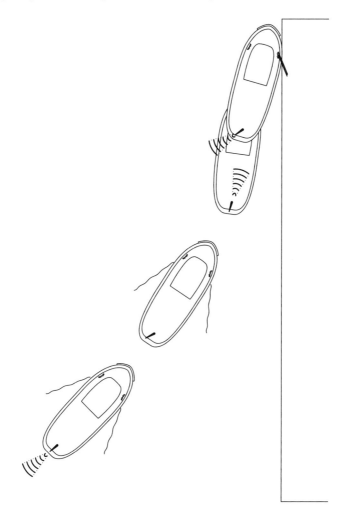

Fig. 7-6. Docking a single-screw tug.

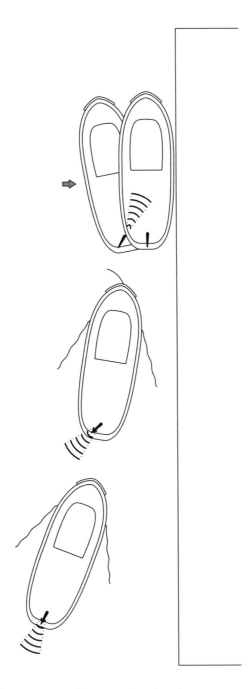

Fig. 7-7. Single-screw tug using torque to advantage.

SINGLE-SCREW WITH FLANKING RUDDERS

Flanking rudders are installed ahead of the propeller (usually two for each propeller) and used to direct the propeller wash when the engines are operated astern. This permits the tug to be steered both when maneuvering astern, and when maneuvering ahead. When maneuvering ahead, the tug fitted with flanking rudders responds much like its conventional counterpart. The flanking rudders are maintained in an amidships position when the tug is operating its engine ahead, so as to prevent the rudders from adversely effecting the tug's handling qualities. If the tug is to be turned in a confined circle (by backing and filling), the flanking rudders are turned in the same direction as the regular rudder when the engine is backed (fig. 7-8).

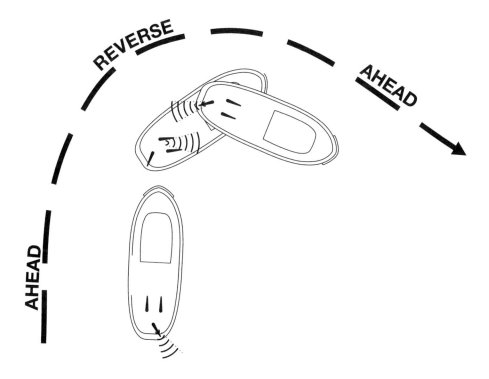

Fig.7-8. Tug with flanking rudders turning around.

SINGLE-SCREW WITH STEERABLE NOZZLE

The single-screw tug with a steerable nozzle has characteristics similar to those of a tug fitted with flanking rudders. It responds to its helm when moving ahead, much as any single-screw tug. The exception is that it may not dead stick (steer with the engine stopped) as well as a conventional tug.

The single-screw tug with a steerable nozzle steers well when maneuvering astern (fig. 7-9). However, in practice, it handles more slowly than a tug with flanking rudders. It requires the engine to be stopped when maneuvering from ahead to astern until the rudder angle is changed (to amidships or reversed) (fig. 7-10). If the engine is not stopped before the helm is changed the tug may swing in the wrong direction.

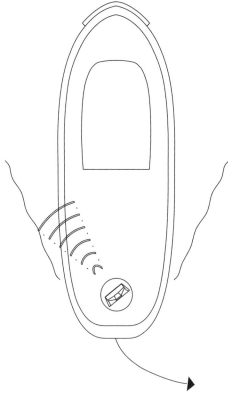

Fig. 7-9. Steerable nozzle tug backing up.

A tug fitted with a steerable nozzle can compensate for the effects of propeller torque in both the ahead and astern directions. This is readily apparent if we repeat the docking approach of fig. 7-7: right hand propeller, starboard side to, in a tug fitted with a steerable nozzle (fig. 7-11).

TWIN-SCREW

Most twin-screw tugs are set up with a right-hand and left-hand propeller. This is to use propeller torque to enhance the tug's ability to turn, steer with engines only, or to move laterally (walk or flank). Placement of these opposite turning propellers can magnify or diminish one or more maneuvering characteristics of the twin-screw tug.

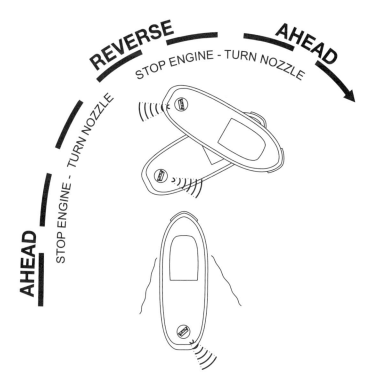

Fig. 7-10. Steerable nozzle tug turning around.

One configuration places the right-hand wheel on the starboard side and the left-hand wheel to port. This is referred to commonly as outward-turning propellers. Torque will add to the effect of each propeller being placed off the tug's centerline. In the case of a right-hand propeller placed on the starboard side the effects of propeller torque (stern to port when backing, bow to port when going ahead) intensifies this effect. A left-hand propeller in this instance diminishes this effect.

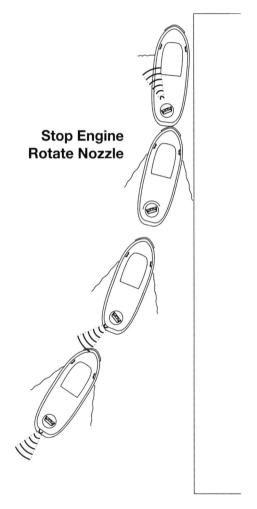

Fig. 7-11. Steerable nozzle tug docking starboard side to.

If the tug is dead in the water and one engine is engaged ahead or astern, with the rudder amidships, the tug responds by moving ahead or astern and turns toward the side of the stopped engine (fig. 7-12). This is caused by the resistance of the stopped propeller and the off-center location of thrust from the other propeller.

Inward-turning propellers have the advantage of better propeller efficiency and enhance the tug's ability to "walk" or move laterally.

Regardless of orientation, both propellers thrusting in the same direction with the same rpm will balance each other's torque and off-center position. By manipulating the balance between the two engines, a twin-screw tug can steer with its engines.

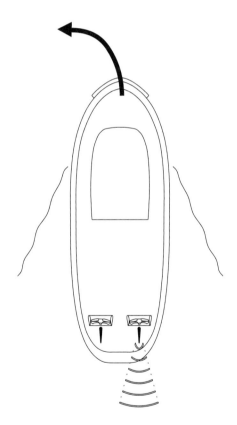

Fig. 7-12. Twin-screw tug effect of one engine maneuvering.

If the tug is dead in the water, with rudder amidships and both engines engaged ahead, the tug will move forward. If the engines are engaged astern, the tug will move back. In both instances, if the speed of the engines is equal, the tug will move in a straight line (fig. 7-13).

To stop or reverse the direction of movement, the engines are simply engaged in the opposite direction until the desired effect is achieved. With the engines engaged ahead, the tug immediately responds to the rudder. The rate of turn is determined by the amount of rudder angle used (degrees) and the engine speed.

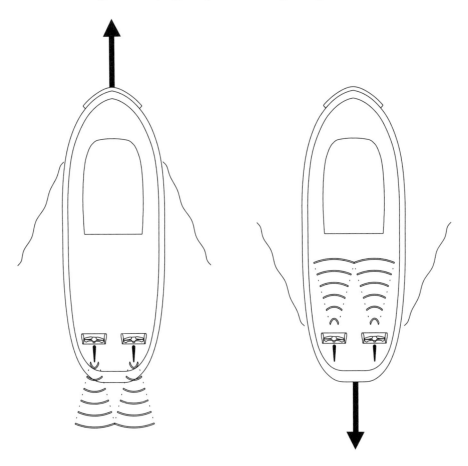

Fig. 7-13. Twin-screw tug effect of balanced rpms coming ahead or astern.

The twin-screw tug shares the single-screw rudder's ineffectiveness when backing until sternway is gathered. When the engines are engaged equally astern, the tug's stern begins to turn in the direction of the rudders only after the tug has sufficient sternway. However, the operator can hasten that effect by raising or lowering the rpm of either engine to induce an imbalance in thrust that begins to turn the tug (fig. 7-14).

None of the situations outlined above take advantage of a twin-screw tug's exceptional maneuverability. To do this, the operator must be proficient in using the engines in opposition to each other. If the tug is going ahead on both engines with rudder amidships, and one engine is backed, the tug turns toward the backing engine and loses speed. If the tug is going astern on two engines, rudder amidships, and one engine is engaged ahead, the stern of the tug is cast toward the side of the propeller turning ahead.

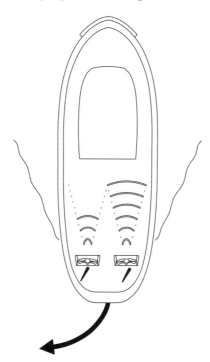

Fig. 7-14. Twin-screw tug effect of imbalanced thrust backing.

This is the technique called twin-screwing (fig. 7-15) Twin-screw left and twin-screw right are terms used to describe the direction the bow moves when this technique is applied. An operator has the option of applying as much rudder angle or engine power as the situation requires. The tug can be made to turn quickly or slowly, and can turn while moving ahead or astern. It can even be turned end-for-end while remaining almost stationary. The variations are endless, but the dynamics are obvious.

"Flanking" is a term used to describe moving a tug sideways. This is a variation of using both engines in opposition (one ahead, one astern) to turn a tug, but in this case the controls are reversed. For example, if the operator wishes to move the tug laterally to starboard, the port engine is engaged ahead, the starboard-engine astern. The rudder is turned to port to make the tug flank to starboard. The effect of the port engine ahead with left rudder normally would turn the boat to port, but backing the starboard engine cancels the turning effect and the movement of the tug ahead.

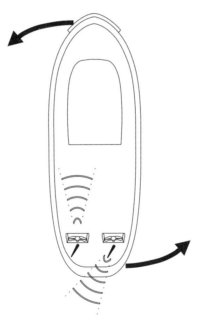

Fig. 7-15. Twin screwing left.

This also is where the advantage of inward-turning propellers comes into play. In the flanking configuration, the paddle wheel effect of both propellers turning in the same direction assists in this maneuver (fig. 7-16).

Undocking a twin-screw tug is a straightforward process. As with a single-screw tug the stern is swung out and the tug then backed clear of the berth. The primary difference is that the twin-screw tug can use its maneuverability instead of the springline required of the single-screw tug (fig. 7-17).

Docking also is a simple process with the twin-screw tug. The operator approaches the dock slowly at a 15° to 20° angle. Nearing the dock the operator can use the twin-screw tug's full capabilities to adjust his approach. At the operator's discretion, the tug can be turned and slowed, turned and speeded up, maintain a straight course, be brought to a stop in a straight line, come to a stop while bringing the stern or bow towards the dock, or any of the variations enabled by manipulating the engine and rudder controls. Once the bow is brought to the dock, a springline can be put up and the stern twisted in (fig. 7-18).

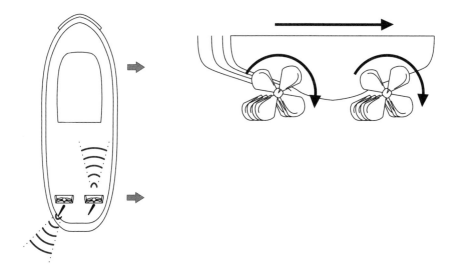

Fig. 7-16. Twin-screw tug flanking to starboard.

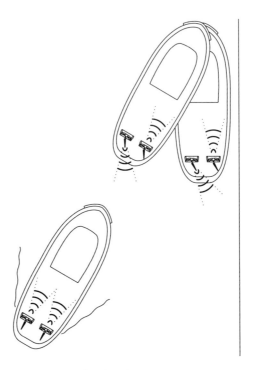

Fig. 7-17. Twin-screw tug departing berth.

It can be seen from the docking and undocking situations described above that flanking is another option of bringing the tug into and out of a berth. Flanking is a maneuver that has its time and place, and in some situations it is the only option. However, this maneuver is prone to overuse. Flanking a conventional tug requires a fair amount of horsepower, causing the tug to create excessive wash in the water, and vibration on the tug. Over time this can create unnecessary wear on the tug's machinery and the port engineer's nerves. Most docking maneuvers have a variety of methods that will produce a successful result. Coming into a berth slowly, giving a few gentle kicks ahead, and then backing easy to stop the tug is the preferred method. However, some operators habitually come in parallel but a good distance off the dock and flank the tug in the remaining distance. While the operator may relish his ability to move the tug sideways, the port engineer may not share his enthusiasm, thinking instead of the unnecessary wear on equipment.

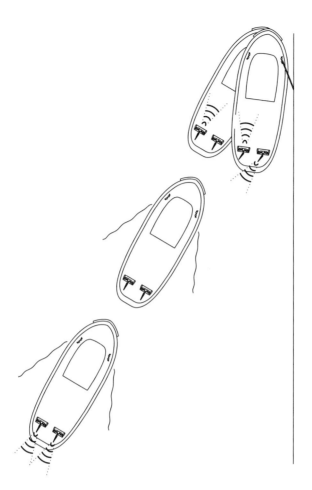

Fig. 7-18. Twin-screw tug docking.

Choosing the appropriate maneuvering option speaks as much to horsepower in the wheelhouse as it does to horsepower in the engine room.

TWIN SCREW WITH FLANKING RUDDERS

The characteristics essentially are the same as those noted above, except that the tug will steer astern on one engine and the tug can be flanked in several different ways.

TRACTOR TUGS

Although tractor tugs can be propelled by either the Voith-Schneider Propeller (VSP) or a steerable propeller, VSPs are more common. This discussion focuses primarily on the VSP configuration.

VSP controls typically consist of pitch levers, a wheel, and an engine rpm control (fig. 7-19). Pitch levers control the magnitude of longitudinal thrust and the wheel controls the magnitude of transverse thrust. Engine rpm is set in accordance to the power demands expected in the maneuvering sequence.

The pitch lever throw is marked in increments of zero to ten, both ahead and astern. Depending on the maneuver or the tow, it is possible to overload the engine if too much pitch is applied. Pitch restrictors can be applied as a precaution to limit the maximum amount of pitch available. The most common causes of engine overload are:

- changing the position of the pitch levers and wheel too quickly
- maneuvering at full pitch (ten) without pitch restriction
- stopping or slowing a vessel with pitch levers set too high
- pulling with pitch levers set above eight
- pushing with pitch levers set above nine

In effect, the wheel moves the bow and the pitch levers move the stern (fig. 7-20 and 7-21).

Because of the tractor tug's maneuverability the operator may find himself operating as much stern first as bow first. The new operator may find himself confused as to which end is which and, more importantly, also may be confused as to the location of the VSP units in relation to the direction of the tug's movement. The VSP units are located at approximately one-third of the tug's length aft of the bow. Many operators prefer to face the "working end" of the tug, no matter which direction the tug is traveling. The VSP controls are very intuitive. Regardless of which direction the tug operator is facing, or which end of the tug is the working end, a left/right turn of the wheel produces a left/right turning motion in the tug.

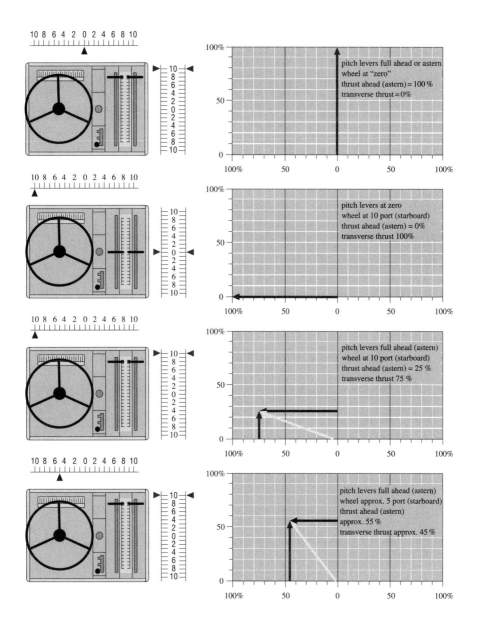

Fig. 7-19. VSP Controls. *(Courtesy of Voith Turbo Schneider Propulsion GmbH & Co.)*

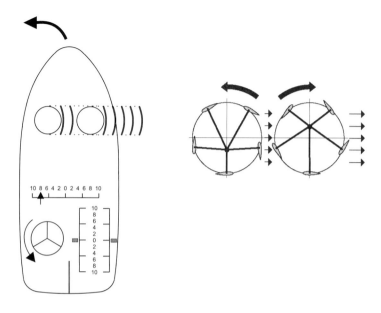

Fig. 7-20. Wheel moves the bow. *(Courtesy of Voith Turbo Schneider Propulsion GmbH & Co.)*

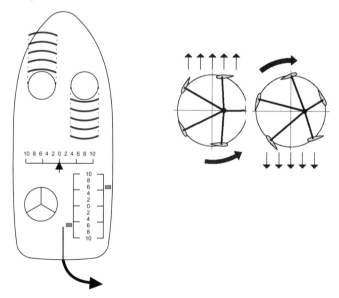

Fig. 7-21. Pitch levers move the stern. *(Courtesy of Voith Turbo Schneider Propulsion GmbH & Co.)*

An important distinction the new operator must recognize is the different turning circles the tug will scribe depending on whether it is operating propulsion-end first or skeg first. When moving bow (propulsion-end) first, the tractor tug will steer into a turn; when moving stern (skeg) first the bow will swing outside the turning circle. The same principle applies to an ASD tug, but the effect is reversed due to the propulsion unit's location at the stern of the ASD tug (fig. 7-22).

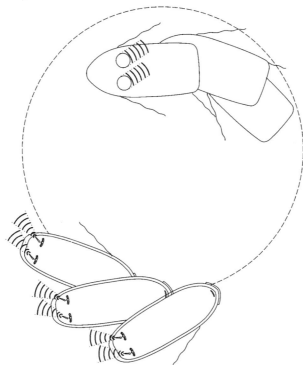

Fig. 7-22. Comparative Turning Effects VSP & Conventional Tugs.

There are two basic methods of stopping a tractor tug. The first is by using the pitch levers only. The second is a combination of using both the pitch levers and wheel. The standard stop is achieved by incrementally pulling the pitch levers back, decreasing and, if necessary, reversing the pitch. This smoothly takes off way and brings the tug to a stop. If a crash stop is necessary, the levers must be steadily pulled to about pitch six of the opposite direction and slowly increased if necessary. This method typically brings the tug to a stop within approximately one tug length.

The second stopping method uses the lateral resistance of the hull. This is accomplished by pulling the levers to 0 and then turning the wheel hard over. The tug will turn 90 degrees from its original heading. The remaining fore or aft motion can be stopped by moving the pitch levers in the appropriate direction. When executed properly, this technique will stop the tug in approximately 0.5 tug length and leave the tug in a position at right angles to its original heading (fig. 7-23).

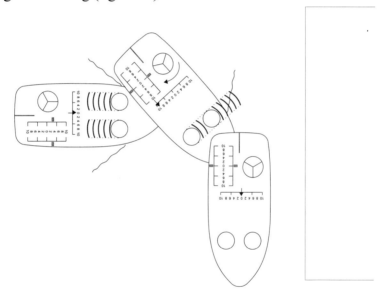

Fig. 7-23. Crash stop. *(Courtesy of Voith Turbo Schneider Propulsion GmbH & Co.)*

A tractor tug, like a conventional twin-screw tug, can be turned in tight quarters in little more than one tug length. However, the tractor tug operator has many more options. He may twin-screw by using opposite pitch settings and leaving the wheel at zero, turn the wheel and use only the transverse thrust, or use any number of combinations of the two. Most operators use the combination technique, enabling fine-tuning of the tug's turning radius and rate, and forward or aft advance to fit the situation at hand. A tractor tug also can be walked or flanked straight sideways. This is done by turning the wheel in the desired direction of lateral movement; and manipulating the pitch levers in a manner similar to flanking a twin-screw tug. In concept, the pitch lever positions are similar to the throttle positions used to flank a conventional tug. The outboard lever (the one closest to the direction of desired movement) is moved to thrust towards the tug's propulsion end and the inboard lever is moved to thrust away from the tug's propulsion end. Fine adjustments between the three controls will propel the tug directly sideways (fig. 7-24).

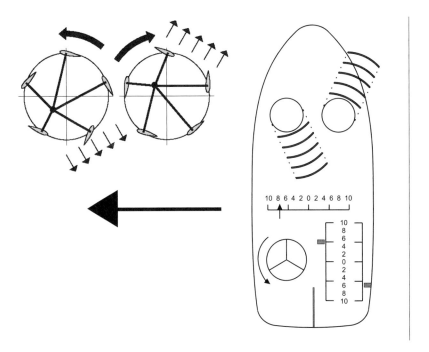

Fig. 7-24. Flanking a tractor tug away from a dock. *(Courtesy of Voith Turbo Schneider Propulsion GmbH & Co.)*

Undocking a tractor tug usually is done bow (VSP) end first. Departing a dock with unrestricted space ahead is a simple matter of turning the wheel away from the dock and pushing the outside pitch lever ahead (fig. 7-25).

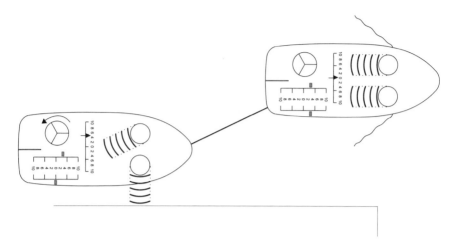

Fig. 7-25. Undocking-Unrestricted Space. *(Courtesy of Voith Turbo Schneider Propulsion GmbH & Co.)*

In more restricted situations the tractor tug can be flanked or walked sideways (fig. 7-24) or brought away from the dock by pivoting the bow (VSP) out and then coming straight away from the berth (fig. 7-26).

A tractor tug can approach a dock either-end first. Regardless of the approach orientation, most operators prefer to secure the stern, (skeg end) first when putting out dock lines. If the tractor tug's bow is secured first, the tug's propulsion end is essentially fixed to the dock, severely limiting the tug's maneuvering leverage. A new tug operator who lacks finesse may work heavily against the bow line in an attempt to move the stern towards the dock. Rather than a tug secured safely at its berth, the result may be the more unfortunate consequence of broken lines and a pulled-up dock cleat.

Stern (skeg) first, the tractor is brought in with the wheel set at "0" and both hands on the levers. The levers are then manipulated to bring the stern into position off the berth and hold it fast. Once secured, the wheel is turned to bring in the bow (VSP) (fig. 7-27).

On a bow-first (VSP) approach, the wheel is turned away from the dock just enough to hold the bow off and bring the tug

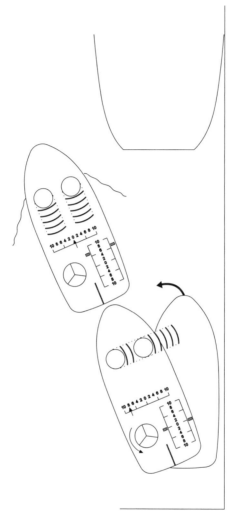

Fig. 7-26. Undocking-restricted space. *(Courtesy of Voith Turbo Schneider Propulsion GmbH & Co.)*

alongside (fig. 7-28). The pitch levers are used to stop any remaining headway. The new tractor tug operator may find that this maneuver requires a higher approach speed than he has experienced with a conventional, stern propelled tug. Without sufficient speed the wheel-induced action of holding the bow off may stop the tug prematurely.

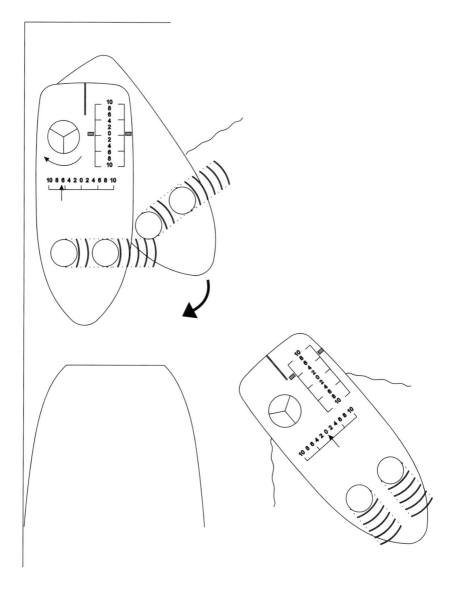

Fig. 7-27. Docking stern (skeg) first. *(Courtesy of Voith Turbo Schneider Propulsion GmbH & Co.)*

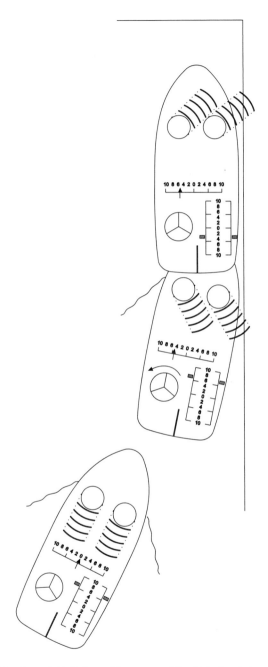

Fig. 7-28. Docking bow (VSP) first. *(Courtesy of Voith Turbo Schneider Propulsion GmbH & Co.)*

Tractor tugs fitted with steerable propeller propulsion units can be maneuvered in the same manner as those with VSP units. Fig. 7-29 shows the basic steering configurations of a steerable propeller tractor tug.

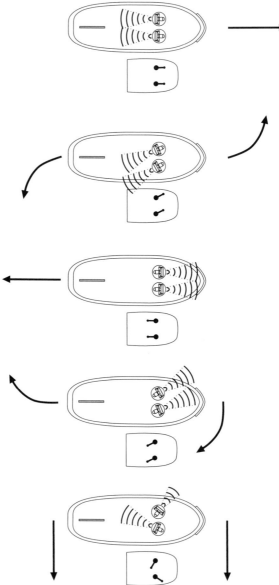

Fig. 7-29. Steering maneuvers with twin steerable propellers.

Some important differences must be kept in mind. Most tug operators who are comfortable with both the VSP and steerable propeller propulsion comment that the VSP controls are more intuitive. Regardless of whether the tug is moving bow first or stern first, an operator turns the wheel in the same direction he wants the tug to go, and pushes or pulls the pitch levers in the direction he wants to direct the resultant of each VSP unit's fore and aft thrust component.

Steerable propeller tugs have a variety of control systems. These range from two separate thruster controls (one for each drive unit), to sophisticated joy stick(s) where computer technology "thinks" for the operator. With separate thruster controls, the direction and magnitude of thrust are contained in one control handle for each drive unit. These separate and independent controls require an operator to be attentive to two tasks; deciding on the appropriate motion for the tug, and how best to configure each drive unit. The joy sticks or similar computer assisted controls relieve the operator of having to conceptualize the drive unit's correct configuration.

Whether in the tractor or ASD configuration, a new operator using separate thrust controls can be easily overwhelmed and confused by the tug's response, for several reasons. First, the tug will not consistently move in the same direction that the control handles are pointed. This is because a control handle points in the direction of the drive unit's thrust, not necessarily in the direction the tug will move. Second, the two units move independently. One can easily enhance, detract, or override the influence of the other. Third, the tug moves and changes direction in accordance with the combined thrust balance of the two units. A slight change in azimuth or rpm in either unit can have a marked effect. The new operator may find himself juggling several variables and must constantly consider the relationship between the drive unit configuration and the tug's actual motion.

With sufficient practice, the steerable-propeller tug operator will find this processing leaves his head and moves into his hands. Maneuvering the tug begins to feel intuitive. This raises the question: if separate thrust controls are so difficult to master why use them at all? Several schools of thought exist concerning this question.

Some feel that only a human can realize the full potential of a steerable propeller tug in the dynamic environment of shipwork. Others feel that computer-aided propulsion configurations provide better tug performance, reliability, and consistency. This chapter will steer clear of this intellectual shoal and focus on the use of separate thrust controls, which provide the best instructional context.

AZIMUTHING STERN DRIVE (ASD) TUGS

ASD tugs do not have a one-to-one relationship between technique and a specific maneuver. All ASD tug motion is determined by the thrust balance between the two drive units. There are three variables for each drive unit: clutched in/out; engine rpm; and azimuth. Combining the range of these variables in two drive units creates a multitude of combinations that may produce the same result. Covering all of these combinations is beyond the scope of this book. The techniques described below are suitable for learning basic ASD tug maneuvers. Once mastered, the operator can begin to incorporate more of the three thrust variables. Only then can both operator and tug reach their full maneuvering potential.

A new ASD operator should remember the following three points. First, he must maneuver the ASD by steering the stern. Second, he must visualize the orientation of the drive units. Third, he must think two to five tug lengths ahead, not only in terms of where the tug will be but how the drive units need to be oriented at that point.

For close quarters maneuvers the operator should keep the drive units in a balanced thrust configuration, where one drive unit is in a position to counteract the other. This provides the quickest response if the dominant thrust direction needs to be reversed. The slowest method of reversing thrust is by rotating the drive unit 180°. This transition can take fifteen to eighteen seconds.

When steering while moving bow first, the drive units are rotated opposite of the tug's desired turning motion. Rotate the drive unit clockwise and the tug turns counter clockwise. This is because the stern is steering the bow. For the bow to go to port, the stern must go to starboard (fig. 7-30).

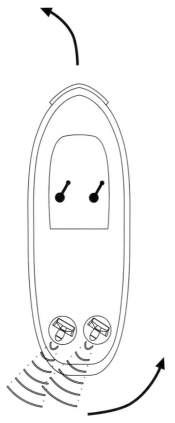

Fig. 7-30. ASD turn to port, bow first.

When steering while moving stern first, the drive units are rotated in the same direction as the tug's desired turning motion (fig. 7-31).

Stopping an ASD can be done either by a transverse arrest, where the two drive units are each thrusting 90 degrees to the tug's centerline in opposition to each other or by rotating the drive units 180 degrees (fig. 7-32).

Flanking or walking is accomplished by starting in a manner similar to a conventional twin-screw tug. The outboard drive unit (one closest towards the direction of lateral movement) is in a

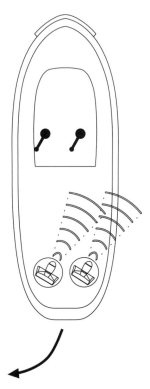

Fig. 7-31. ASD turn to port, stern first.

backing alignment, while the inboard drive unit is angled ahead and in (fig. 7-33). The azimuth and rpm of each drive unit are adjusted to produce the desired effect. There is an important difference between flanking or walking a tractor and ASD—the ASD cannot produce lateral motion with the same speed and strength of a tractor tug.

An operator requires awareness of three additional principles when learning how to dock and undock an ASD tug. The first is that he has the option of employing several different methods to accomplish the same maneuver. The second is to choose the technique appropriate to his experience and skill level. The third is to always leave an out, that is, always have the units configured in such a way that he can check the tug's motion toward the dock or other obstacles. The techniques described below are basic techniques. As the new operator gains experience and skill, he can employ more advanced methods.

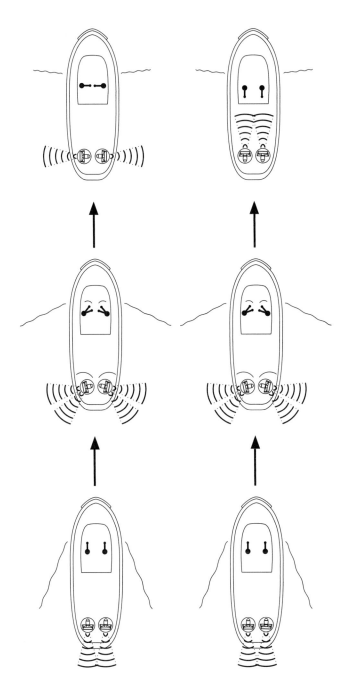

Fig. 7-32. ASD stopping methods.

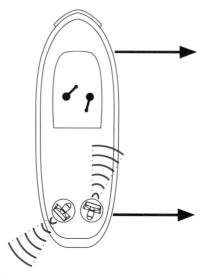

Fig. 7-33. ASD flanking.

An ASD usually undocks from a berth stern first. In an unrestricted space the outboard drive unit can be clutched in at 90 degrees to lift the stern off and then rotated aft to bring the tug away from the berth. The off-center position of the drive unit brings the bow off the dock and the azimuth can be adjusted to lift the stern at an equal rate while the tug is moving aft (fig. 7-34).

In restricted space the tug can align the inshore unit at 90 degrees and the offshore unit at 180 degrees. Clutching in the inboard unit lifts the stern off the dock. Clutching in the outboard unit begins to move the tug aft and checks the swing of the bow toward the dock (fig. 7-35).

When docking, the operator slowly swings both units towards a partial transverse arrest as he approaches the dock. This slows the tug down and still allows the operator to keep the intended heading. As the tug's bow nears the dock both units are brought to the full arrest position, or slightly aft, to stop and hold the bow while a line is passed over. Once the bow is secure, the operator can clutch out on the inboard and kick the stern over with the outboard drive unit. The inboard unit is in a position to check the stern if necessary (fig. 7-36).

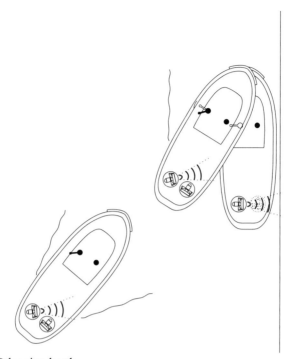

Fig. 7-34. ASD leaving berth.

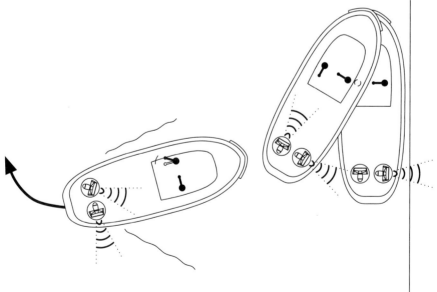

Fig. 7-35. ASD leaving berth.

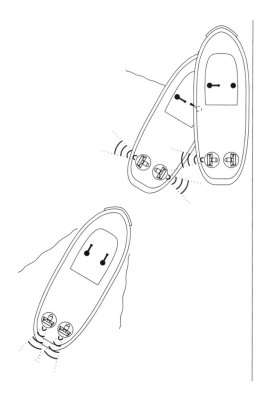

Fig. 7-36. ASD docking.

One common problem with this method is that the inboard unit's propeller wash is directed toward the dock. A dock with a solid face tends to keep the stern off and make the bow dive for the dock prematurely. An alternate method can be employed to prevent this.

The operator can come in on the same approach but bring the tug to a stop about one tug width off the dock. At this point he can rotate the two units into a walk toward the dock. The walking configuration avoids direct propeller wash towards the dock and gives the operator both fore and aft, and lateral control of the tug (fig. 7-37).

Regardless of tug type, once an operator is competent handling the light tug he is ready to apply those skills to the next level of basic tug handling for ship work.

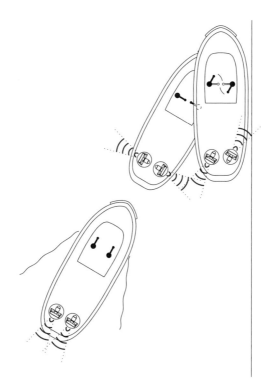

Fig. 7-37. ASD docking.

STUDY QUESTIONS

1. What is a critical decision point?
2. Describe the effect of propeller torque on conventional single-screw tugs with a right-hand propeller.
3. How is a single-screw tug maneuvered when it is necessary to back for some distance?
4. How is a single-screw tug turned in a short distance?
5. Will a single-screw tug turn more readily in one direction than the other?
6. What is the principal advantage of flanking rudders or a steerable nozzle rudder on a single-screw tug?
7. Are steerable nozzles as effective as flanking rudders?

8. How does a twin-screw tug steer with engines?

9. What is meant by the term twin screwing?

10. What is meant by the term "flanking," and how is it done?

11. How do conventional tugs usually undock?

12. How does a conventional tug normally dock?

13. What do the wheel and pitch levers control on a VSP tractor tug?

14. A tractor tug will steer _____ a turn.

15. What two methods can be used to bring a VSP tractor tug to a stop?

16. How can you flank or walk a VSP tractor tug?

17. Why is it sometimes necessary to have a higher approach speed when docking a VSP tractor tug?

18. Do the control handles on an ASD tug always point in the direction of the tug's motion?

19. How can you flank or walk an ASD tug?

20. Why must you think ahead on an ASD tug?

CHAPTER EIGHT

SHIP-TUG INTERACTION AND TUG HANDLING

Shipwork requires the highest level of tug handling skill. The operator's decisions are made in a multidimensional context that must accommodate the relative motion of the ship, the hydrodynamic interaction between ship and tug, and the direction and amount of force to be applied to the ship. Multitasking at this level requires a foundation of light tug handling experience and a solid understanding of the factors influencing the interaction between ship and tug.

When both a tug and ship are in motion, tug handling becomes an exercise in relative movement. From the tug handler's perspective, the most obvious characteristic is the *appearance* of ship motion and speed, which can be deceptive. However, both tug operator and ship handler must remain aware of a much more subtle, but equally important form of relativity: the allocation of the tug's available horsepower.

Regardless of design, a tug has a fixed amount of thrust capability. Thrust capability is dictated by its engine horsepower, propulsion and steering mechanism, and hull shape. When underway there is always a proportional relationship in how this capability is allocated. In any maneuvering situation the tug must balance three factors: its own hull resistance in the water, opposing forces when maneuvering to the desired position, and application of force to the ship by pulling or pushing. Tugs are most effective when these three factors are working in unison, as when a tug tows a slowly moving ship in a straight direction. A tug running straight ahead at full power allocates all its thrust to one direction. As soon as the helm is turned some of this thrust is expended to turn the tug, overcome the tug's increased lateral resistance, and continue moving forward. As more rudder angle is applied, the tug encounters more lateral resistance and more thrust will be taken away from its ahead component. As a result, the thrust devoted to moving the tug ahead diminishes, causing the tug to slow down as it turns. Pulling or pushing on a ship may further reduce the thrust allocation for each component.

Both pilot and tug operator must be aware of how circumstances dictate and limit the choices of thrust allocation. If a ship is moving at nine knots, the tug needs to apportion most of its thrust just to keep pace with the ship. Little thrust will be left for the tug to steer and even less to apply to a towline. This can become critical when the tug is trying to counteract the effects of hydrodynamic interaction.

The hydrodynamic interaction between ship and tug is a function of ship speed, displacement, and hull shape. As a ship moves through the water it creates different pressure zones and water flow velocities around its hull (fig. 8-1). In simple terms, a ship moving bow first is displacing or pushing water out of its way, in essence, making room for the body of the hull to pass. Water resists this displacement and builds up at the ship's bow. This results in a high-pressure zone that pushes floating objects, including tugs, away from the bow.

Relative to the ship, the water flows faster near the ship's hull than farther away. This increase in water speed creates a low-pressure zone by the side of the hull. The same principle applies to the tug and when the two are running in close proximity, side by side, the tug and ship are pulled together.

As the water nears the ship's stern it comes under the influence of the propeller. This accelerates the water flow and accentuates the drop in pressure, increasing the suction towards the ship's stern. Just aft of the propeller the wash and ship's wake generate another small area of higher pressure. Generally, this area of high pressure is not as pronounced as the one encountered near the bow.

These are the same principles that create the bow cushion and bank suction effect in deep draft ships when they transit shallow, narrow channels. The exact locations of higher and lower pressures depend on ship hull shape and whether it is in ballast or loaded.

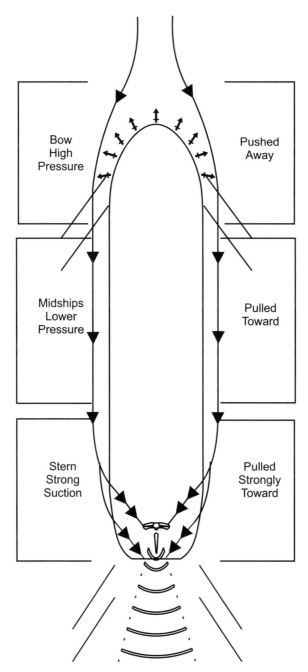

Fig. 8-1. Ship hull pressure zones.

COMING ALONGSIDE

A tug coming alongside a moving ship is influenced by the respective pressure zones of both ship and tug. In general, the tug is pushed off at the bow and experiences increasing suction toward the ship's stern. These interaction forces can be extremely powerful and dangerous if overlooked by either the ship pilot or tug operator.

Ship speed is a major factor in the amount of ship/tug interaction. A ship's hull resistance increases as the square of its speed. Increasing speed from four to six knots doubles hydrodynamic resistance. Accordingly, the high- and low-pressure zones, and their respective repelling and suction forces, increase exponentially. The ship pilot must be vigilant about his speed when working with tugs alongside his ship. A small change in speed may appear insignificant from the ship's bridge, but at the waterline it can have large consequences. The tug operator also should be aware of the ship's speed, since it will help him anticipate the magnitude of the different pressure zones.

Tugs working alongside moving ships maneuver in and out of the different pressure zones. Many times they may be in transition areas where one end of the tug is under the effect of high pressure and the other under lower pressure.

Coming alongside a ship's aft quarter (fig. 8-2), the tug's first encounter with turbulence and higher pressure is created by the ship's propeller wash and wake. The tug operator may have to increase power to push through the wake. As the tug's bow gets past the wake and breaks the plane of the ships transom, it comes under the influence of the ship's suction zone at the stern. Since the tug's bow is closer to this suction than the tug's stern, the tug's bow turns toward the ship. This is the first transition zone.

The tug operator must immediately correct by steering the bow away from the ship and power past the tumble or overhang on the stern. As the tug continues to progress forward and comes abeam of the aft quarter, the operator must increase engine speed to keep pace with the accelerated water flow and counter the strong suction forces towards the ship.

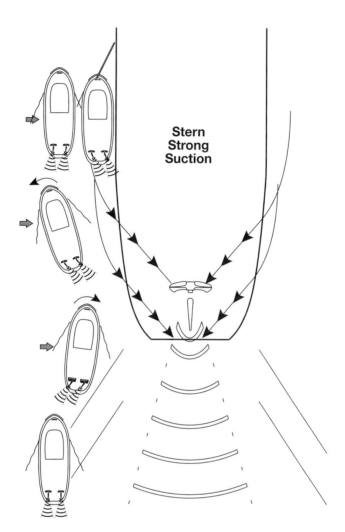

Fig. 8-2. Conventional tug coming alongside a ship's quarter.

Once past the stern rake and running parallel to the flat of the ship the tug enters yet another transition zone. The tug's own high pressure at the bow pushes it away from the ship, while the stern continues to be sucked toward the ship. Again, the tug operator must respond quickly. Once the tug is on a stable heading, paralleling the ship's course, and is abeam of the flat shell of the ship, it can come in alongside.

The tug operator must be aware of four hazards when attempting to come alongside a ship's quarter. First, suction towards the ship tends to be most powerful near the stern. Second, if the tug comes alongside the ship prematurely, the rake or tumble of the stern may cause serious structural damage to both tug and ship. Third, the ship's turning propeller presents a constant and ominous hazard. Fourth, the ship may change course or be swinging in a turn as the tug comes alongside.

These suction forces at the stern can be surprisingly strong and very unforgiving but also can cause serious accidents. One of the worst scenarios may occur when a tug operator who, while bringing his tug alongside, does not anticipate or react soon enough to the tug's bow being pulled toward the ship's stern. His efforts to correct may be too little or too late and the tug may end up underneath the ship's counter. The ship's suction can hold the tug like a powerful magnet. The tug operator may apply more power and rudder in an attempt to break the bow free, but this action only exacerbates the situation. As the ship's suction acts on the underbody of the tug, the ship's counter pushes on the tug's deckhouse. These forces act like a vertical lever, heeling the tug over to the point where the outboard side of the tug's deck becomes immersed in flowing water. At this point, the tug loses all maneuverability and may slide down the ship's counter in a semi-capsized state to the propeller.

A more controlled method of bringing a conventional tug alongside the aft quarter of a moving ship is to pick a safe place to land clear of the ship's flare or counter, stand off the ship a few tug lengths, shape a course parallel to the ship, and adjust speed to match that of the ship. Usually, pacing the ship with the tug's helm turned slightly away from it compensates for the ship's suction, and allows the tug to be gently eased in alongside. As soon as the tug comes alongside, the helm can be put over toward the ship with the engines running ahead to keep the tug in position until the lines are fast.

Ideally, the tug should contact the ship about midships on the tug. The bow can then be gently swung in toward the ship as the helm is put over. In cases where ship suction is very strong, one option is landing the tug farther forward where there will be less suction, and then letting the tug slide back into position. Some ship captains may grumble about this "messing up the paint job," but scuffed paint is preferable to a dent.

One of the true advantages of a tractor or ASD tug becomes apparent when coming alongside a moving ship. A tractor operating VSP first, or an ASD operating stern first, is more maneuverable and safer because the propulsion units are on the up current end of the water flowing by the tug. In this configuration, the operator controls the leading end of the tug and counters the forces of ship interaction by steering toward or away from the ship (fig. 8-3A).

This is particularly advantageous on ships such as container ships that have long flair at the bow and long counters at the stern. These types of ships may have a flat side shell on only the center one-third of the vessel. Many times the only available chocks for push/pull are located at the bow, or stern, or mid-body of the ship. A tractor or ASD tug can safely come in close aboard the aft quarter, pass a towline across, and then either run parallel to the ship with a slack line, or move up the ship to the flat side shell (fig. 8-3B). Once the ship slows and is ready to be breasted to the dock the tug can swing out perpendicular to the ship and position itself to push/pull.

A tug coming alongside the midships section of the ship is drawn toward the ship even though the water flow is more uniform. In addition to this suction, the high pressure at the tug's bow can force its bow away from the ship. To steady the bow a conventional tug uses its rudder. A tractor or ASD tug operating propulsion end first uses its propulsion to direct its leading end toward the ship's hull.

Although this is considered one of the safer locations to come alongside a ship, it is not without hazard. A typical application of this maneuver is picking up or dropping off a pilot. It sometime happens that after the pilot leaves the bridge an overenthusiastic bridge team may prematurely start building speed and altering course before the pilot has disembarked. This increase in speed magnifies the effects of ship suction. The turning and advance of the ship's hull, particularly if the tug is on the outside aspect of the ship's turning circle, may trap the tug alongside. The tug may be able to free itself only if the ship immediately takes to a steady heading, reduces speed, or both.

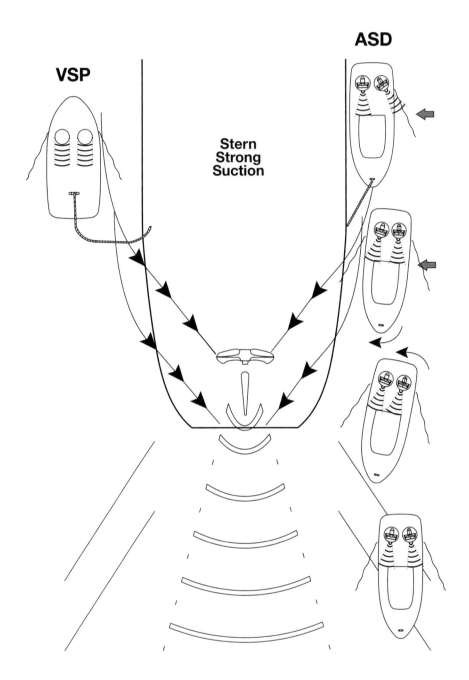

Fig. 8-3. Tractor, ASD tugs coming alongside stern quarter.

PASSING A TOWLINE TO A SHIP'S BOW

Putting up a towline on the bow of a moving ship will quicken the heart rate of a tug operator, whether new or experienced. This maneuver requires intense concentration and a deft touch on the throttles and steering controls. A mistake in judgment or tug handling in these circumstances can be disastrous. Working a towline on the bow of the ship involves four phases: initial approach, passing the towline, towing on the line, and casting off.

A tug operator has to balance several factors when deciding on his initial approach to pass a towline. Ship speed, bow shape, and tug capability all play important roles in the decision.

In general, the high-pressure zone at the ship's bow pushes the tug away from the ship. The location and intensity of this zone varies widely depending on ship speed, bow shape, and draft. The more fine and slender the bow profile, the farther aft and less pronounced the pressure zone (fig. 8-4). *But with deeper draft and fuller bow shapes, the stronger the pressure zone.*

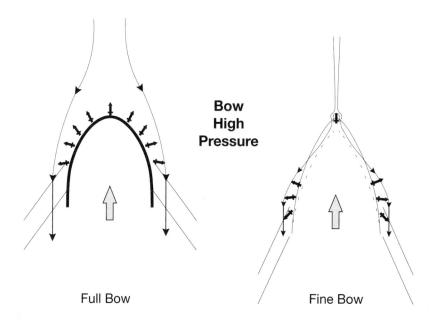

Fig. 8-4. Pressure zones, different bow shapes.

The safest initial approach is for the tug to position itself well forward of the ship and allow the ship to come to the tug (fig. 8-5A). This approach is unsettling to many ship captains and pilots as they lose site of the tug well before it is underneath the ship's bow and the tug is directly in the oncoming ship's path. Another approach is to come in from a position off to the side and forward of the ship where the tug remains in sight of the bridge (Figure 8-5B). This approach minimizes the time spent directly in front of the ship during the approach and passing the towline phases. With either approach the objective is for the tug to assume a position parallel and in front of the ship, but advancing at a slower pace allowing the ship to slowly close the distance. Once the tug is within 2-3 tug lengths of the ship, the tug should pace the ship and get a feel for the effect of the ship's pressure wave (Figure 8-5C).

At this point the tug is still in a good position to abort the maneuver in the event of a misjudgment or mechanical failure. The prudent tug operator may use this safe zone to let the tug settle down and make sure that he has a good feel for how the tug is interacting with the ship. This also is a good time to make certain the deck crews on the ship and tug are ready with the requisite heaving lines, messenger lines, and boat hooks to pass the towline. Once all parties are confirmed ready, the operator can bring the tug into the best position to pass the towline safely.

This position can vary depending on the shape of the ship's bow. Ships with fine bows tend to have less disturbed water directly underneath the bow. However, ships with bulbous bows still produce a marked pressure wave in front of the bulb. In addition, there may be a low pressure "hole" just aft of the bulb. The tug can come into a position and hold right underneath the bow overhang or slightly off to one side (fig. 8-5D). This may not be an option with a blunt-bowed ship, such as a bulker or loaded tanker. The bow pressure wave on these vessels can extend quite a bit forward of the stem, which may require the tug to hold off center (fig. 8-5E). In this position the tug may experience a turning motion towards the center line of the ship, due to the pressure wave pushing on the end of the tug closest to the ship. Additional factors to consider may be the direction of the wind or if the ship is turning. Generally it is safer to be on the leeward side of the bow. The wind will blow the heaving or tag line away from the ship, allowing the tug to hold a little farther off the bow. If the ship is turning, a position on the inside of the turn will help avoid hitting the bulb.

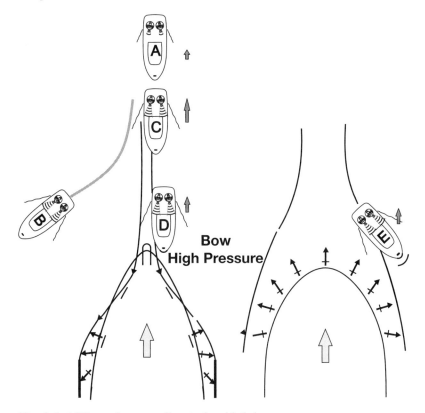

Fig. 8-5. ASD passing a towline to the ship's bow.

Once in position to pass a towline, the tug and its crew are in a vulnerable position. The tug is operating close to a ship's overhang, stem, and, in some cases, bulbous bow. A tug's maneuverability can be lost immediately on even the slightest contact with any of these structures. The tug may be hard to hold steady, as its hull may be in different areas of pressure, and the tug's propulsion wash may be hitting the ship's hull. The deck crew is working on the exposed deck, passing heaving and messenger lines while the ship's bow looms overhead. A dropped line fouling the propeller or a line coming taut prematurely may place the crew and tug in immediate peril. The tug operator must give his full attention to the task at hand. Inattentiveness or an error in judgment can have serious consequences.

Once the towline is passed the tug pays out an appropriate length and moves into position to follow the pilot's directions.

Sometimes circumstances dictate that the tug approach from aft of the ship's bow (fig. 8-6A). In this approach the tug overtakes the ship while standing off a safe distance. Once in this position (fig. 8-6B), the tug slows to pace the ship and moves into position to pass the towline (fig. 8-6C). With the towline slack and made fast to the ship, the tug moves into a position to tow (fig. 8-6D).

A tug approaching the ship's bow from aft may transition through several different pressure zones. The operator must guard against "stemming." Stemming occurs when, due to improper maneuvering, the tug ends up crosswise in front of the ship, being pushed by the ship's stem.

One common cause of stemming is misjudging the tug's set toward the ship's bow on initial approach. The tug operator may be angling in, steering for a point forward of the ship's bow. The accelerated water flow just aft of the ship's bow may slow the tug down relative to the ship. If the operator does not sense and react immediately to the loss of relative forward speed, the tug can set in more rapidly toward the ship's flare.

On a conventional tug, the tug operator's instinct is to counteract the increased set toward the ship with more power and rudder to take the tug's stern toward, and the bow away from the ship. If he has left enough room between the tug and ship, he can recover. If not, the tug's stern quarter can contact the ship and pin the tug against the bow. When more power is applied, the tug may slide down along the ship's bow until the tug comes broadside and is run down (fig. 8-7A).

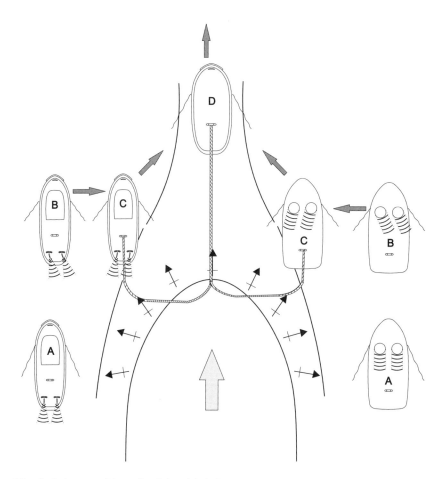

Fig. 8-6. Approaching aft of the ship's bow.

In this position the only recourse for the tug operator may be to back full, damaging the tug and ship, but this is better than being run down by the ship, holed, or capsized.

A second common cause of stemming is over or under steering. A tug running parallel to the ship close aboard requires more power to push through the ship's bow wave as it nears the ship's forward shoulder. Once the tug's bow has broken through, it moves into the high-pressure zone around the bow. This pushes the bow of the tug away from the ship. If unanticipated, this push may appear sudden and forceful to the tug operator. He may counteract instinctively with a large amount of rudder to bring the tug's bow back towards the ship. However, as the tug continues to move into the high-pressure zone, the stern may be pushed away from the ship. If the operator is not attentive and does not immediately swing his rudder the other way, the combination of excess rudder and interaction with the stern can cause the tug to sheer rapidly toward the pathway of the oncoming ship and be run over (fig. 8-7B).

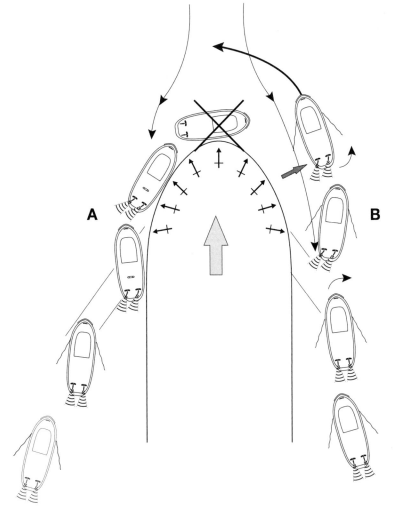

Fig. 8-7. Stemming situations.

This does not imply the tug operator has sole responsibility for passing a towline successfully. The ship pilot also has a crucial role. The pilot controls one of the most critical factors in this maneuver: ship speed. A ship speed of four to seven knots is considered a good guideline for passing a towline safely. This maneuver can be conducted at higher speeds, seven to ten knots, but success requires three critical elements: the tug must have specialized design features, the operator must possess an in-depth familiarity with the tug and extensive experience in the effects of ship interaction.

At high speeds the design differences between tugs becomes critical. A tug operator must be well versed in the capabilities and limits of the specific tug he is operating. Even sister ships may have crucial differences in horsepower, propeller pitch, underwater profile, and control systems, each of which may affect performance. Even experienced tug operators have had serious stemming accidents while operating a new tug.

Casting off the towline is as demanding a maneuver as passing it over. The operator must hold the tug underneath the bow while the ship crew pulls enough slack to remove the eye from the bitt and pass the line back to the tug. The tug's deck crew must be vigilant in retrieving slack or dropped lines before the lines get fouled in the tug's propulsion.

The principal hazards of passing a towline near the bow apply to all tug designs. However, tractor tugs operating bow first, and ASD tugs operating stern first, have distinct safety advantages over conventional tugs. They operate propulsion-end first and can extricate themselves from a potential stemming situation by steering away from the ship. The propulsion end of the tug moves away from the ship and pulls the remaining length of the tug to safety (fig. 8-8). Even if the trailing end of the tug (stern in the case of a tractor tug, bow in the case of an ASD operating stern first) strikes the ship, maneuverability is not lost as with a conventional tug. The propulsion units are "up current" from the ship's bow, which places them in less disturbed water. They also are at greater distance from any lines dropped in the water off the ship's bow.

Of course, if an ASD tug is operating bow first, it shares the same risks as a conventional tug.

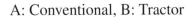

A: Conventional, B: Tractor

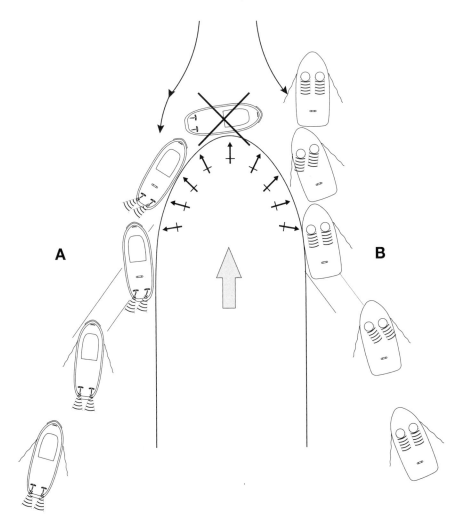

Fig. 8-8. Comparative advantage of tractor/ASD vs. conventional tug in stemming situations.

When casting off, ship speed should be the same as when the towline was passed. Most ships have a considerable lag time between the issuance of an engine order and its actual manifestation at the waterline. An inexperienced pilot may order an increase in ship speed while releasing assist tugs. He may assume the tug will retrieve its towline and be clear before the ship actually begins to speed up. This can be problematic if there is any line handling or mechanical problem that adds time to the casting off process. The ship's crew may not be present to cast off, towlines can foul in ships anchors, winches can jam, and operators can pay out too much slack, allowing the towline to wrap around the stem or bulbous bow.

All of these problems can occur while the tug operator is holding the tug in one of the most precarious positions in shipwork. If the ship begins to increase speed while these problems are being sorted out, it raises the risk to the tug and the blood pressure of the operator. Since the ship's interactive forces grow exponentially with the ship's speed—the bow pressure wave becomes more powerful. The ship speed may approach and surpass the tug's maximum hull speed, overwhelming the tug. Speed is as important a factor casting off as in connecting the towline.

PASSING A TOWLINE TO A SHIP'S STERN

Passing a towline to the stern of a ship is similar to putting one up on the bow: initial approach, passing the towline, towing on the line, and casting off. However, the ship's wake-and-propeller wash introduces an additional layer of complexity. Aft of the ship's transom is a mixing zone of propeller wash, high-pressure, and suction zones. The water is turbulent and the boundaries between these different zones may not be apparent. The water in the ship's propeller wash may be moving two to three knots faster than the ship's wake outside the wash zone. This exponentially magnifies the hydrodynamic forces acting on the tug's skeg or keel.

In a good initial approach the tug overtakes the ship, towing point first, on a course just outside of the most disturbed water. For a tractor tug this is stern first. For an ASD or conventional tug putting a towline off the bow, this is bow first. Since the pattern of the ship's disturbed water is fan-shaped, the tug can move closer as it overtakes the ship (fig. 8-9 A).

Once the leading end of the tug is even with the plane of the ship's transom, the tug should slow down and pace the ship parallel to its heading (fig. 8-9B). This is the position in which the operator lets the tug settle down and ensures that the ship and tug crews are ready to pass the towline. Many ships have an extensive suction area around the ship's stern. Forward of the transom plane, this suction moves the tug laterally toward the stern flare or tumble. However, the tug operator can minimize these hazards by staying aft of the transom plane and off center. Unfortunately, the beam and shape of many ship's sterns are not conducive to the tug remaining off center to pass a line to the centerline stern chock. In this event the tug must maneuver directly in the ship's propeller wash and hold close off the ship's stern. Ship pilots can make this process safer by stopping the ship's propeller while the tug hooks up. Tug operators can assist in the process by pushing on the ship's transom directly under the chock if the height of the tug and the shape of the ship's stern allow.

Once the line is made fast to the ship, the tug operator can pay out line until the tug is in position to take the pilot's directions. If the line is dropped in the water the operator should immediately steer the propulsion units away from the line.

As in passing a towline to the bow, the ship pilot has an important role to play. Ideally, ship speed is seven knots or less and the ship's propeller (if fixed pitch) is stopped while the tug is making fast. If this is not possible it is helpful if the pilot can maintain a consistent amount of turns and rudder while the tug is passing the towline. This gives the tug operator a chance to acquire a feel for the boundaries and strength of the ship's wash and suction zones. If the pilot must make a change in rudder or engines, particularly reversing the engines, he should communicate this to the tug operator. Many ships

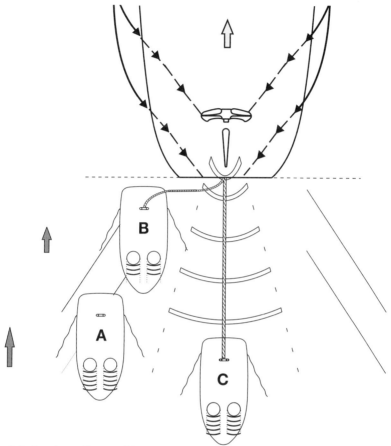

Fig. 8-9. Passing a line to ship's stern. A: Tug approaching stern. B: Holding in safe position. C: Ready to work.

when first engaging the propeller astern create a deep trough close to the ship's stern that will rapidly pull the tug into the transom.

The primary hazard in casting off from the ship's stern is a dropped or slack towline. Attentive winch operation is called for when shortening the towline. Holding the tug in a position close to the ship's chock will minimize this risk. It is always prudent, if the ship will not be put at risk, to stop the propeller when casting off the tug. Particular vigilance should be exercised when casting off from a ship with sternway. A dropped line under those circumstances can easily foul the ship's wheel.

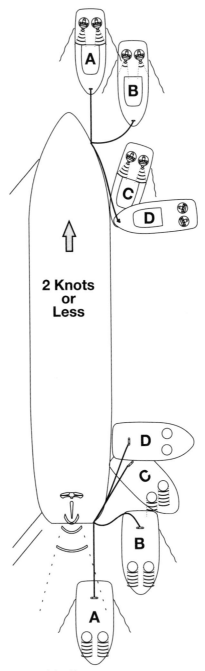

Fig. 8-10. Transitioning to push/pull.

TRANSITIONING FROM TOWING TO PUSH/PULL

An additional advantage of tractor and ASD tugs is their ability to transition from towing to a push/pull position without disconnecting the towline (fig. 8-10). General practice consists of waiting for a ship to slow to two knots or less before transitioning to a push/pull position. The tug operator then moves the tug outside the ship's side shell and runs a course parallel to the ship. The towline is adjusted to allow the tug to move down the ship's side to the desired position. Once the towline becomes taut, the tug pivots out to push/pull.

Once a tug operator has safely maneuvered a tug into position and made fast to the ship, he can declare to the pilot the tug is "in position and ready to work." Working the tug requires the same amount of skill and finesses as it did to get in position. The next chapter addresses the principles of working tugs effectively in four basic shipwork positions.

STUDY QUESTIONS

1. What three factors must a tug operator balance when maneuvering around ships?
2. Hydrodynamic interaction between ship and tug are a function of which three factors?
3. What hydrodynamic pressure zones are caused by the ship's passage through the water?
4. What are the hydraulic effects on the tug as it makes an approach close to a ship underway?
5. How should the tug handler compensate for these effects?
6. What is stemming?
7. What are two common cases of stemming?
8. Should a tug ever back its engine when coming alongside a ship?

CHAPTER NINE

BASIC TUG POSITIONS IN SHIPWORK

Tugs assisting ships use four basic positions:

1. Leading tug towing on a line
2. Tug alongside the ship pulling on a line or pushing
3. Tug alongside the ship breasted or "on the hip" towing
4. Trailing tug, pulling on a line or pushing

Of course, these positions have many variations. Pilots and tug captains have developed innovative modifications of these positions to suit the unique environmental, geographic, and berthing conditions of various ports. Over time these variations may become the custom at different ports. However, the modifications and variations are founded on the fundamental application of pushing and towline forces associated with the four basic positions.

For present purposes the terms lead tug and tail tug refer to positions relative to the ship's direction, not its bow or stern. As an example, the lead tug tows off the bow of a ship moving bow first, and the stern of a ship moving stern first.

UNDERSTANDING PUSHING AND TOWLINE FORCES

Effective application of a tug pushing, pulling, or working a towline can be achieved only if both pilot and tug operator understand the dynamic balance between a tug's center of hydrodynamic pressure, lift-and-drag forces, maneuvering and turning resistance levers, and applied horsepower (fig. 9-1).

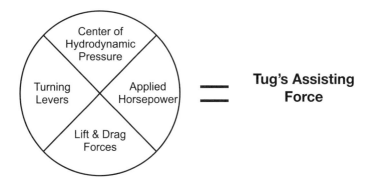

Fig. 9-1. Balance of towline force.

Center of hydrodynamic pressure

Water flowing around the tug's hull generates hydrodynamic forces. The source of this water flow may be a tug propelling its hull through the water, a tug being dragged through the water by a taut line, propeller wash from either the tug or ship, or any combination of these. Water flowing around a tug's hull generates both lift and drag forces. The amount of lift and drag is related to the angle between the hull and the direction of the oncoming water flow. This is similar to the "angle of attack" that provides lift in an airplane wing. The culminative effect of these hydrodynamic forces can be represented by a figurative point on the tug's hull: the center of hydrodynamic pressure (CHP) (fig. 9-2).

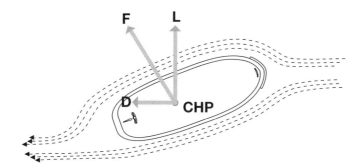

Fig. 9-2. Center of Hydrodynamic Pressure (CHP). D: Drag Component. L: Lift Component. F: Resultant Hydrodynamic force.

The CHP is neither a fixed location nor a constant magnitude. It migrates forward or aft as a function of the tug's hull shape and angle of water flow. Speed affects its magnitude. Just as in ship interaction, a tug's hydrodynamic resistance increases as the square of velocity.

An important consequence of a tug's hydrodynamic resistance is it also can produce powerful beneficial forces. These forces can be transformed into either a pushing force into the ship or a pulling force on a line. The effect of water flow acting on a tug's hull can be broken down into two major components: one is a drag force that parallels the direction of water flow, the other is a lift force.

In simple terms, a skilled operator can use the tug's hull to enhance its pushing or pulling force. The hull's hydrodynamic resistance effectively magnifies the weight of the tug's hull and, when applied correctly, may proportionally increase the tug's pushing or pulling force. This is the force that makes indirect towing so effective. It can also have beneficial effects in conventional towline and push/pull applications.

The paradox is that this hydrodynamic resistance, so beneficial in one set of circumstances, can be a detriment in another. The conventional tug towing on a line as the lead tug uses this effect to advantage. Yet if the same tug tows on a line as a tail tug, hydrodynamic resistance severely limits the tug's maneuverability. The capability of a tug to manage this effect is inherent to the tug's design. A primary design factor in determining this capability is the longitudinal turning lever created by the distance between the tug's center of hydrodynamic pressure and towing point.

The interrelation between a tug's center of hydrodynamic pressure, lift & drag force, and maneuvering and turning levers varies as a function of the type of tug and the ship assist technique employed.

LEADING TUG TOWING ON A LINE

As shiphandler and tug operator begin to understand pushing and towline forces, it becomes apparent that towing on a line as the lead tug is more complex than simply pointing the tug in the desired direction of pull. A common misconception among those unfamiliar with towline work is that the tug's heading must always parallel the desired direction of towline pull to be effective. This is not only incorrect but can result in the tug quickly girting or tripping. The ship does not necessarily respond to the direction in which the tug is steering. The ship always responds to the angle and amount of force on the towline.

To perform this task effectively, the operator must balance three tasks. First, he must position the tug so that the towline angle, *as measured between the ship's heading and the direction the towline is tending from the ship to the tug's towing point,* is at an appropriate angle. The greater the angle the more sideways component to the pull.

Second, the operator must apply force to the towline. The amount of force applied to the towline is a function of the tug's thrust and steering forces combined with hydrodynamic lift and drag forces.

Third, the operator must ensure that the tug's forward motion keeps up with the advance of the ship.

These tasks can best be illustrated in the example of a tug moving from an in-line position to one 30 degrees off center of a ship moving ahead at four to five knots. Fig. 9-3 shows this on a conventional tug.

Conventional Tug

In fig. 9-3 the conventional tug operator gives a little left rudder to cant the tug's bow at a slight angle to the incoming water flow. Once the flow has caught the bow, the tug operator moves the rudder closer to midships as the water flow pushes the tug off to the side. The operator can regulate the speed of the tug's offset movement by manipulating the tug's angle to the incoming water flow. The operator must exercise caution, since the greater the angle, the greater the loss of the tug's forward motion due to increased hydrodynamic resistance. The tug operator may have to compensate for this by applying more thrust.

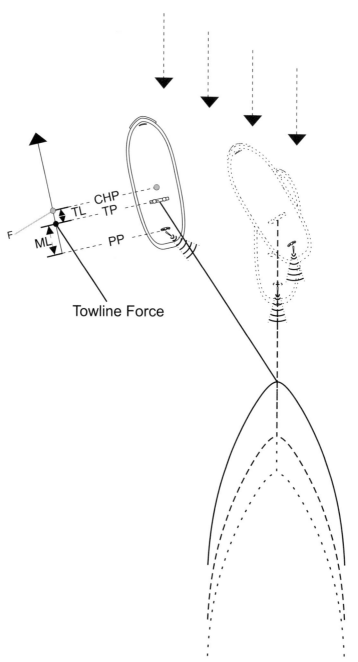

Fig. 9-3. Conventional tug maneuvering on towline. CHP to TP = Turning Lever (TL), TP to PP = Maneuvering Lever (ML)

Once the tug has moved to a position where its towline angle approaches the desired angle (30 degrees), additional forces come into play.

To understand these forces, visualize two levers simultaneously at work. The first is a turning lever (TL). This is the lever formed by the distance between the tug's center of hydrodynamic pressure (CHP) and the towing point (TP). Since the tug is at an angle to the oncoming water flow, the CHP migrates to a point close to amidships just forward of the towing point. When the tug is in this position, the incoming water flow applies a force turning the bow to port, while the towline pulls the stern to starboard.

This is counteracted by the tug's longest maneuvering lever (ML), the distance between the tug's propulsion point (PP) and towing point. When the operator applies a little right rudder and increases thrust at the tug's propulsion point, it counters the turn by moving the tug's stern and towing point away from the ship. This adds to the tug's pull on the towline. In effect, the towline is tensioned by the tug's turning force and the hull's hydrodynamic resistance.

Slight rudder and throttle adjustments keeps the tug in this position, pulling on the towline, but advancing forward with the ship. In practice, these forces are not calculated, they are sensed. One of the key sensory cues is the amount the tug is heeling.

When the towline is leading off at an angle to the tug's centerline, the tug heels in proportion to the force being exerted on it and the angle of the towline. The stronger the force, the greater the heeling angle. One perception associated with an increase in heeling angle is the tug operator's feeling of the tug "digging into" the water. This is a sign that the hull's hydrodynamic forces are strengthening.

When the tug heels over, the operator adjusts his stance to maintain his balance on the tilted deck. This is the same sensory tool the operator may use driving the tug, i.e., using the sense of tilt to gauge the balance between the tug's forward advance, towline angle, and pull.

This balance point can be elusive and right on the limit of the tug's capability. As a result, when the ship pilot calls for a pull off the ship's centerline he may see the tug constantly making slight changes in course. The tug operator is searching for that balance point and wary of surpassing it. Again, the degree of heel is the clearest indication of when that limit has been passed. When the tug reaches the point of deck edge immersion it is a clear sign that the tug is on or past the precipice of losing its maneuverability. It may have lost its ability to safely manage the influences of towline force, hydrodynamic drag, and thrust forces. At this point the towline must be either slacked or released, since the tug may quickly trip or girt.

Conventional tugs are very effective in the position of the lead tug on a towline. The tug is capable of applying forces equally to both the port and starboard directions. The tug can shift from side to side unencumbered by the ship's flair. Changing steering directions requires the tug to shift to a position that has the new, desired towline angle. Tug response to changes in ship steering orders is not immediate, and depends on how quickly the tug can shift laterally to the new position.

In addition, when a conventional tug on a towline begins a turn, the bow of the ship tends to be set initially in the direction opposite to the intended turn. This is due to the tug rotating its propulsion point around its towing point before pushing the towing point to a new position. The initial turn moves the towing point opposite to the direction of the intended turn. This phenomenon will last until the towing point has shifted laterally to the point the towline angle, as measured from the ship's bow, leads in the direction of the desired turn. This effect is most apparent on small or light draft vessels and less noticeable on a heavily laden ship. If anticipated, this tendency can be useful. Tug skippers use this effect all the time and often "snap" their barges by applying power in a turn to swing them clear of obstacles (fig. 9-4).

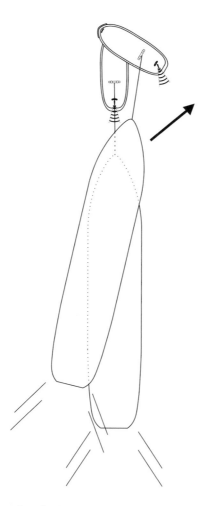

Fig. 9-4. Effect of snapping the tow.

Tractor or ASD-Reverse Tractor Tug

This section uses the term tractor tug, but the principles apply
equally to an ASD tug operating as a reverse tractor. A tractor tug
moves off the ship's centerline by steering its propulsion end in the
desired direction of pull (fig. 9-5). Because the tug's propulsion
units are up-current from the towing point, the tug essentially must
propel itself into position instead of letting hydrodynamic forces
set the tug sideways.

Once in line at the desired angle, the tractor tug operator must balance the same forces as those in a conventional tug. However, the interplay and relationship between the forces are very different. This is because the propulsion point is located ahead of the towing point and center of hydrodynamic pressure. As with a conventional tug, the force of the incoming water and the pull on the towline exerts a turning effect on the tug. In this case (fig. 9-5) the tractor tug's leading and trailing ends are rotated counterclockwise around the towing point.

However, unlike the conventional tug, when a tractor tug applies its maneuvering lever to counter the turning force, the propulsion point moves toward the ship's centerline. Because the propulsion point is oriented up-current from the towing point and center of hydrodynamic pressure, the tug will exert proportionately less force on the towline.

This orientation does not have the same hydrodynamic advantage as a conventional tug in this position. Most of a tractor tug's hydrodynamic resistance transforms into a drag force, not a lift force. To exert effective sideways force on the ship's bow, the tractor tug must be more in line with the direction of the towline pull. Unlike a conventional tug, an increase in towline angle requires the tug to present a more broadside aspect to the incoming water flow. As a result, the drag on the hull increases. The more drag on the hull, the more the propulsion capability has to be allocated to pull the tug sideways through the water. Consequently, less force is applied to the towline.

This presents a bit of a paradox to the tug operator. One might assume that the greater the towline angle, the greater sideways force will be exerted on the ship. But this is not necessarily the case, particularly as ship speed increases. As ship speed and towline angle increase, the tractor tug must devote more thrust to pulling its own hull through the water, reducing the actual pull on the towline.

When shiphandling with tugs, sometimes what you see may not be what you get. For example, when a ship is making headway of six to eight knots, and the pilot calls for a tractor tug to give "more pull to

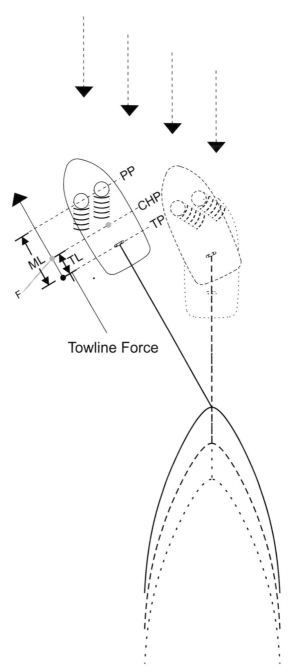

Fig. 9-5. Tractor (ASD) tug maneuvering on towline.

starboard," the pilot's eyes may equate an increase in towline angle with an increase in sideways pull. However, the reality may be very different. The tug operator may be using all his skills and tug capability to prevent being passed and overrun by the ship. Relatively little force actually is applied to the towline.

Design differences between tractor and ASD tugs may affect an ASD's capability to work as a lead tug in the reverse tractor configuration (bow-to-bow). An ASD tug operating bow-to-bow has the advantage of a longer maneuvering lever, but the disadvantage of operating skeg first.

The distance between the towing point and the propulsion point on an ASD is nearly the entire length of the tug. This produces a more effective maneuvering lever. However, the skeg on an ASD, particularly if it extends well aft, may detract from the tug's effectiveness. When operating as a reverse tractor, the skeg may add drag to the aft end of the tug, reducing lateral mobility. In addition, if the skeg extends well aft, the propeller wash may hit the skeg, adding both unwanted drag to the hull, and reduced thrust effectiveness (fig. 9-6).

When working as a lead tug on a line, both a tractor tug and an ASD reverse tractor have distinct safety advantages over a conventional tug. The propulsion-first orientation of these tugs, combined with omni-directional thrust capability, make them less vulnerable to girting and tripping situations.

However, the effectiveness of all tugs working as a lead tug on a line diminishes as ship speed increases. Although the conventional tug may be able to generate more lift forces with its hull, it will increasingly become more limited in its ability to pull at high towline angles off center of the ship. The tractor and ASD tug, although able to tolerate higher towline angles longer, will devote more and more of their bollard pull to overcome the tug's increased drag through the water.

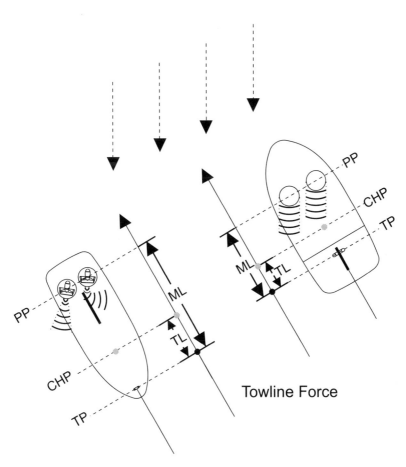

Fig. 9-6. Tractor/ASD Reverse tractor comparison. Note longer maneuvering lever of ASD reverse tractor.

TUG ALONGSIDE THE SHIP PULLING ON A LINE OR PUSHING

Tugs working alongside may make fast to the ship with one, two, or three lines. The more maneuverable the tug, the fewer lines required. A single-screw conventional tug will typically put up a headline, a springline, and a quarter line (fig. 9-7).

Springlines on tugs doing shipwork usually are led from as far forward as practical on the tug, which is the bullnose or bow chock. This permits the tug to pivot freely. If a springline is made fast farther aft by the forward shoulder bitt, the springline keeps the tug from

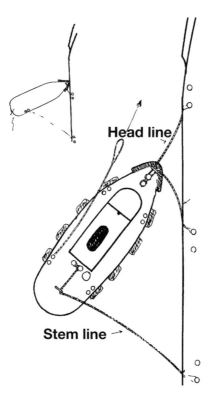

Fig. 9-7. Two methods of making a conventional tug rest alongside.

turning as much as it may need to get into good position (45 degrees to 90 degrees to the ship's centerline). The headline, also referred to as the backing line, is sometimes made fast farther aft if both lines will not fit through the tug's bow chock

Stern lines or quarter lines are interchangeable terms. They refer to lines that are led from the after quarter bitt, or stern of the tug to a position on the ship that maintains the tug approximately at right angles to the ship. This is done to counteract the effect of the torque of the tug's propeller or the effects of the ship's movement ahead or astern.

Quarter lines also are used to counteract the effect of wind and current. It is important to remember that when the quarter line or the stern line is in use, excessive way on the ship is dangerous as this can part the line or even capsize the tug.

The more maneuverable twin-screw, tractor, and ASD tugs can eliminate the need for spring and quarter lines as circumstances permit.

The magnitude of a tug's pushing force against a ship's side is a result of the aggregate effect of four factors:

- Pushing angle of the tug in relation to the fore and aft line of the ship
- Amount of the tug's hydrodynamic lift
- Amount of the tug's thrust allocated for steering
- Efficiency of the tug's maneuvering leverage

The pushing angle of a tug has both a transverse and longitudinal component relative to the ship. The transverse component pushes the ship laterally and the longitudinal component adds to the ship's speed of advance. The magnitude of the tug's push is a result of both the tug's thrust and the amount of lift generated by the hydrodynamic resistance of the tug's hull. If the ship is moving through the water and the tug is at an angle to the oncoming water flow, its hull will generate both lift and drag forces (fig. 9-8).

A portion of the tug's thrust must be allocated to a steering force to overcome the tug's hydrodynamic drag and keep the tug at the appropriate angled position. A key determinant to the required allocation of steering force is the proportional relationship between two levers. The first is the tug's maneuvering lever (ML): the distance between the towing point (TP) and the tug's propulsion point (PP). The second is a turning lever (TL): the distance between the towing point (TP) and the center of hydrodynamic pressure (CHP) (fig. 9-9). The longer the maneuvering lever is in relation to the turning lever, the less thrust is required to keep the tug in position and the more is available for pushing.

A good illustration of this relationship is a tug breasting a ship off a berth and continuing to push while the ship comes ahead to gain steerageway (fig. 9-10). When a tug is pushing on a ship dead in the water, the towing point and center of hydrodynamic pressure are both at the bow of the tug. The tug can apply full power in the

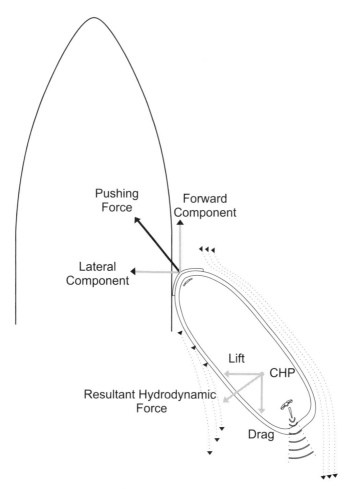

Fig. 9-8. Force components of a tug's push.

direction of the ship since the turning resistance lever is minimal. The tug has no problem keeping 90 degrees to the ship (fig. 9-10A). As the ship begins to move forward, the tug's hull begins to resist the broadside water flow. The turning resistance lever lengthens as the CHP moves aft (fig. 9-10B). The tug has to apply a combination of steering and additional power to counteract the increased leverage of water flow. In effect, the tug allocates some of its forward thrust to steering to hold the tug in position.

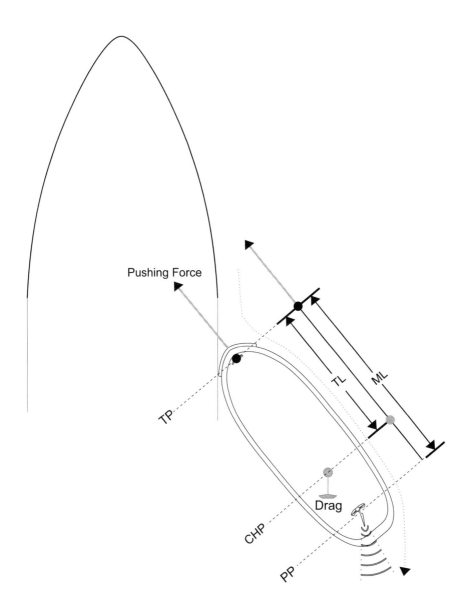

Fig. 9-9. Maneuvering and turning levers of a conventional tug pushing.

Even though the tug has had to divert some of its thrust away from its primary purpose of pushing directly into the ship, it remains effective. This is due to the lifting force created by the tug's hull resistance. As ship speed and the tug's lateral resistance increase, more of the tug's capability has to be used to counter the drag force of the water flow.

Eventually, the forces of hydrodynamic resistance overcome the tug's steering and propulsion capability and the tug begins to fold in parallel to the ship's side (fig. 9-10C). This threshold primarily is a function of tug design. A conventional tug's center of hydrodynamic pressure generally is farther aft than on tractor or ASD tug. This creates a longer and stronger turning lever that must be overcome for the tug to stay in position. A conventional tug can put up a quarter line to assist in holding up the stern, but generally a ship speed of three to four knots is considered maximum for conventional tugs to push effectively.

ASD and tractor tugs can maintain effective pushing forces at higher speeds. This is due to their omni-directional thrust capability, a CHP located farther forward than on a conventional tug, and the lift produced by the hull's resistance (fig. 9-11).

These same principles apply when a tug is alongside a ship pulling on a line (fig. 9-12). When a tug is pulling on a stationary ship the tug can allocate all of its thrust to pulling on the line. As in the pushing example, when the ship begins to gather some headway, the tug's center of hydrodynamic pressure migrates from a location near the towing point to one closer to amidships. The conventional tug quickly loses efficiency backing on a headline at any but the lowest speeds. This is because it lacks maneuverability astern to overcome the drag component of hydrodynamic forces. Unless a quarter line is up, the tug quickly folds in alongside the ship's hull. The ASD and tractor tug can maintain their 90 degree aspect by using the dual advantage of their omni-directional thrust capability and their shorter turning lever. In addition, the ASD and tractor tug can "load" their towlines with the lift force of their hull's water resistance.

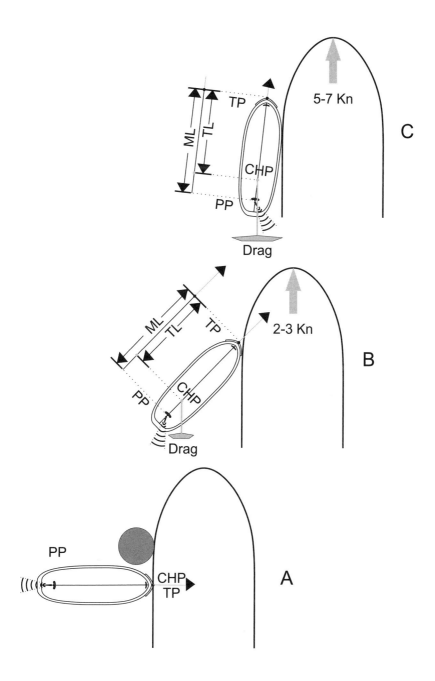

Fig. 9-10. Conventional tug pushing on ship.

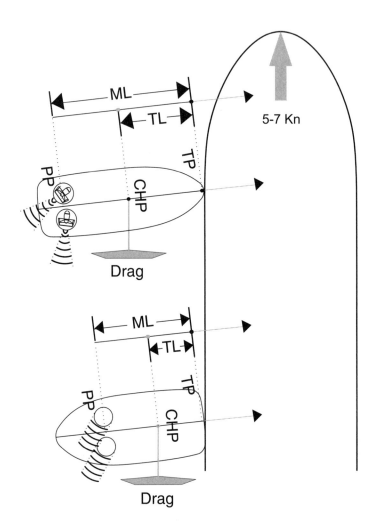

Fig. 9-11. Tractor and ASD tugs pushing on ship.

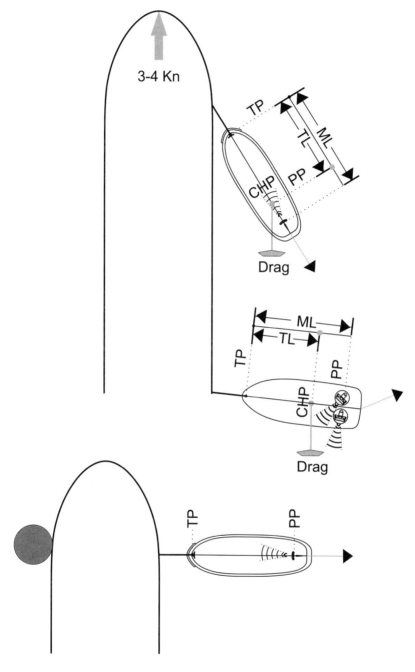

Fig. 9-12. Conventional and ASD tugs backing on a line.

One added performance factor when pulling on a line is the distance between the tug's propulsion units and the ship's side. Too short a distance will cause the tug's propeller wash to push on the ship's side. This may reduce or completely eliminate the desired effect of pulling on the towline. This is why tractor and ASD tugs orient themselves with towing point toward and propulsion units away from the ship's side in push/pull configurations. Extending line length may minimize this effect.

TUG ALONGSIDE THE SHIP BREASTED, OR ON-THE-HIP TOWING

Breasted or on-the-hip towing are terms used to indicate a tug is solidly lashed alongside a vessel. The tug is secured alongside a barge or ship, usually with a minimum of three lines: headline, springline, and stern line.

It is important that the tug be secured parallel to the ship's centerline with its propulsion units as close to the end of the ship as possible (fig. 9-13). This gives the tug better leverage in applying steering forces to the ship.

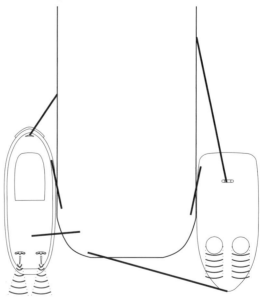

Fig. 9-13. Conventional and tractor tugs secured for breasted towing alongside a ship.

The objective of alongside towing is to make the tug a temporary, but integral, part of the ship. In this position the tug provides both steering and propulsion to the ship. It can assist in propelling the ship forward and backward, and provide steering assistance.

Maintaining the tug's parallel aspect to the ship's centerline is key when made up alongside for breasted towing. If the lines are too slack, the tug's angle to the ship's centerline varies and detracts from the tug's ability to apply effective steering forces.

When a tug is made up for breasted towing the ship responds to the tug's applied force through the lines and the tug's hull-to-hull contact with the ship. The tug relies on tight lines to leverage its hull and thrust against the ship. Slack lines allow too much variance in the tug's alignment which can reduce its effective leverage.

As an example, a tug with tight lines, made up on the port quarter of a ship, can turn the ship to port by working its propulsion ahead with left rudder (Fig. 9-14A). If the tug has a slack headline, it will "toe out" under the same application of thrust and rudder (Fig. 9-14B). The spring line is now pulling the ship ahead and towards the tug. The tug's hull is still pushing on the hull but because the tug is "toed out" the angle of the tug's thrust has changed in relation to the ship's centerline. Its thrust is less lateral and more forward. The result is that the ship accelerates ahead with minimal turning.

Although it sounds good in theory to spot the tug solidly at the end of the ship, this often proves difficult in practice. Most tugs can make up only alongside the flat of the ship's hull or in areas of minimal flair or tumble. There are two reasons for this. The first is to avoid contact between the superstructure of the tug and hull of the ship. The second is that the tug is more secure and less likely to yaw on the lines if more surface area near the tug's propulsion end is in contact with the ship. Tugs making up alongside ships with long flairs and counters may find the only suitable locations are hundreds of feet from either end of the ship.

Ship chock placement also can have an effect on where a tug can make up for breasted towing. Vessels such as car carriers, cruise ships, and container ships may have only a few chocks on deck for

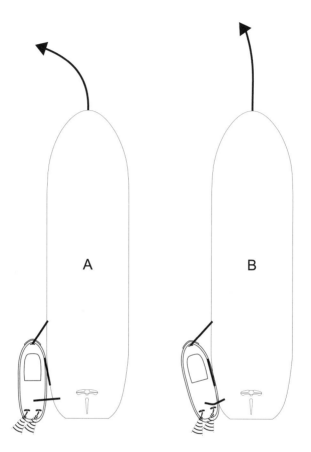

Fig. 9-14. Effect of tug's angle in breasted towing.

making up the tug and these may be badly placed. Many have their chocks and bitts located so far forward or aft that the tug must be extremely careful to avoid damage.

If the tug cannot position itself so its propulsion end is past the end of the ship, the tug's propulsion units may lie only a few feet from the ship's hull. When the tug thrusts away from the ship, the tug's wash will interact with the ship's hull and may result in two detrimental consequences. First, it may negate the desired turning effect on the ship and second, it may induce a heeling angle towards the ship.

Conventional, tractor, and ASD tugs are all well-suited for breasted towing. The omni-directional thrust capabilities give the tractor and ASD tug an advantage over the conventional tug.

TRAILING TUG PULLING ON LINES OR PUSHING

Conventional Tugs

A conventional tug functioning as a trailing tug can work either off its stern with a towline or off its bow with one or two headlines. Working a towline off the tug's stern is effective at low ship speeds and when a braking and steering effect is desired. A conventional tug fares better working headlines off the tug's bow when circumstances call for more ship speed.

When a conventional tug operates as the trailing tug on a towline, the tug usually is dragged backwards through the water. This places its center of hydrodynamic pressure well aft of the tug's towing point, and close to its propulsion point (fig. 9-15A). This is a very unstable and dangerous position. The center of hydrodynamic pressure (CHP) now exerts significant leverage on the tug's stern due to its distance from the towing point. High amounts of thrust and steering force are required to counteract this lever due to the limited maneuverability of conventional tugs when backing up. A conventional tug easily can be overwhelmed by this force at anything but the slowest ship speed.

This risk can be managed by mechanically moving the towing point through the use of hold downs and a gob rope. Proper use of a gob rope moves the towing point over the center of hydrodynamic pressure. This balances the tug's turning lever, allowing the tug to maintain a broadside aspect at very slow ship speeds (fig. 9-15B). The tug can now exert steering forces on the ship by coming ahead or astern.

When the tug provides only a braking force to the ship, it is safer to shift the towing point all the way aft to a position over the propulsion point (fig. 9-15C). The change in the location of the towing point alters the tug's handling characteristics in two ways. First, the tug tends to be pulled through the water stern first by the

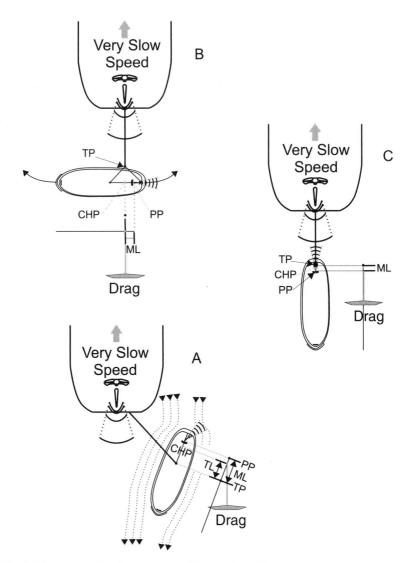

Fig. 9-15. Conventional tug on a towline as the trailing tug.

ship and stays in line with the ship's heading, minimizing the risk of capsizing. Second, it greatly reduces the effectiveness of the maneuvering lever. The minimal horizontal distance between the towing point and the propulsion point severely compromises the tug's ability to steer. It is a safer, but less maneuverable, configuration.

When circumstances call for more ship speed, a conventional tug is safer running bow first as the trailing tug. In this position, the tug assists ship steering by functioning as a "rudder" tug. As the tug turns and presents more broadside to the water flow, its hull generates hydrodynamic resistance and lift forces similar to a rudder. This rudder force can be accentuated by applying more of the tug's power in the direction of lift. These forces are transferred to the ship through the tug's headlines, the tug pushing on the ship's hull or a combination of the two (fig. 9-16).

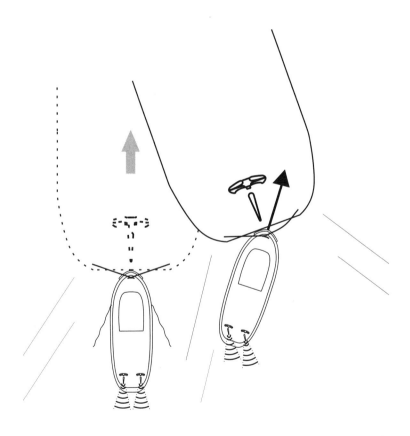

Fig. 9-16. Conventional tug on split headlines as the trailing tug.

When operating as a trailing tug off the ship's stern, line configuration and placement of the tug is determined largely by the ship's stern shape, chock placement, and draft. A loaded ship with a flat transom may allow the tug to push on the ship's stern. A short headline close to the ship's centerline will do in this case.

If the ship is in ballast or has a canoe stern the tug may not be able to push on the hull and must rely solely on the lines to transfer its applied force. A split headline works well in this situation.

A split head consists of two lines run from the bow of the tug, one to each side of the ship's stern. In this configuration when the tug's bow is steered to starboard, the ship turns to port and vice versa. When the tug runs its engines astern it adds an in-line braking effect to the ship. The tug operator should be aware that the line load is greatly increased when backing on split headlines due to the sharp line angle. As noted previously, the ship pilot should communicate clearly to the tug operator when the ship will be operating astern propulsion. Powerful ship propellers can suck the tug rapidly toward the ship's stern when first operating astern. This risk can be managed easily with good communication between pilot and tug, allowing the tug to anticipate this effect.

A conventional tug can be very effective functioning as trailing tug off the bow of a ship with sternway. The tug can apply force on either side of the stem. It does this by shifting from one side of the bow to the other by putting up a "wrapped headline." The wrapped headline is a line that leads from one side of the ship's bow to the bullnose of the tug located on the opposite side of the ship (fig. 9-17). The tug turns at the appropriate angle either to push on the bow in one direction or push into the line for the other.

As long as the tug can stay in position by working ahead it can respond quickly to a pilot's order for a change in steering direction. However, bulbous bows, severely raked stems, and ships with flare may prevent a tug from working in this position.

Tractor and ASD Tugs

As a trailing or rudder tug, tractor and ASD tugs have the option of using two techniques to apply steering forces: direct and indirect

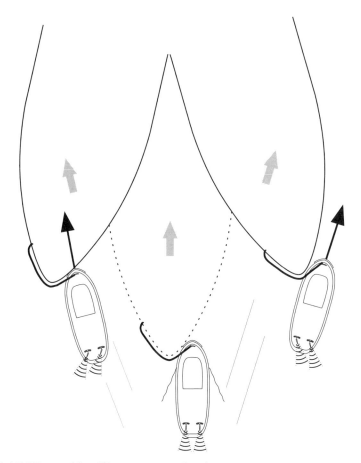

Fig. 9-17. Wrapped headline on conventional tug.

towing. Both types of tugs operate with their towing points forward and propulsion points aft. A tractor will be stern to stern; an ASD is bow to stern. (This section of the book uses the term tractor tug, but the principles apply equally to an ASD tug operating as a reverse tractor.)

Direct Towing

In direct towing the tug pulls on the line in the desired direction (fig. 9-18A). The limits of this configuration are the same as a tractor or ASD pulling on a line in the push/pull configuration (fig. 9-12). Generally, at ship speeds of five to six knots, direct towing is an effective means of applying steering forces. Unless the tug has special design features, it rapidly loses its effectiveness at ship speeds of more than six knots. As ship speed increases, more of the tug's thrust must be allocated to counter the increased drag force on the tug's hull.

Indirect Towing

Unlike direct towing, indirect towing (fig. 9-18B) uses the tug's hydrodynamic resistance to add, rather than subtract, force applied to the towline. The tug operator establishes the desired towline angle by driving the towing point off to the side. In this position the angle of attack of the oncoming water flow creates lift and drag.

The hydrodynamic drag and the pull of the towline seek to turn the tug in line with the direction of water flow. The tug can counter this turning force with its maneuvering lever, the distance between its towing point (TP) and propulsion point (PP). As the operator applies this corrective action to maintain the tug's angled aspect to the incoming water flow, he transfers the forces generated by hydrodynamic resistance and the tug's maneuvering lever to the towline. As ship speed increases, hydrodynamic resistance increases, and subsequently the towline forces increase.

A variation of indirect towing is powered indirect towing (fig. 9-18C). In this technique, the tug operator adds to indirect towline forces by moving the tug's towing point farther off to the ship's side and applying full power. Moving the towing point causes the towline angle to approach 90 degrees in relation to the ship's centerline. Applying full power provides additional thrust to the hydrodynamic and maneuvering forces already being applied to the towline (fig. 9-19). At a five to seven knot ship speed, this action has the potential to apply 75 to 120 percent of the tug's rated bollard pull to the towline.

The maximum ship speed at which a tug can safely function in indirect towing is a function of the tug's design. One speed threshold does not apply to all tractor or ASD tugs. A tug's stability, freeboard, bollard pull, skeg design, and towing and propulsion point locations are all determinant factors in establishing safe limits.

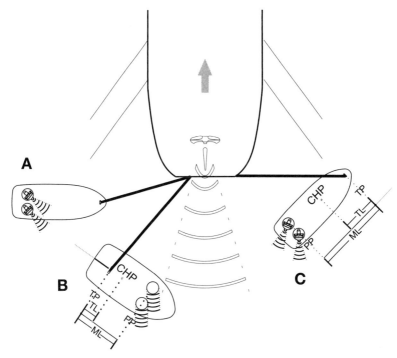

Fig. 9-18. Direct, in-direct, powered in-direct towing modes.

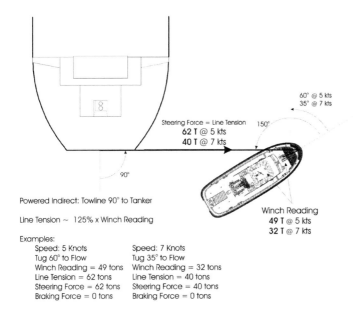

Steering Force = Line Tension
62 T @ 5 kts
40 T @ 7 kts

60° @ 5 kts
35° @ 7 kts

150°

90°

Winch Reading
49 T @ 5 kts
32 T @ 7 kts

Powered Indirect: Towline 90° to Tanker

Line Tension ~ 125% x Winch Reading

Examples:

Speed: 5 Knots	Speed: 7 Knots
Tug 60° to Flow	Tug 35° to Flow
Winch Reading = 49 tons	Winch Reading = 32 tons
Line Tension = 62 tons	Line Tension = 40 tons
Steering Force = 62 tons	Steering Force = 40 tons
Braking Force = 0 tons	Braking Force = 0 tons

Fig. 9-19. Powered indirect towline forces. *(Courtesy of Glosten & Associates.)*

The forces generated by indirect towing are the result of a combination of several factors: towline angle in relation to the ship, tug angle in relation to the towline, speed through the water, and allocation of the tug's thrust to steering. These multiple variables make it difficult to gauge the actual force being applied to the towline. In such situations a tensiometer can help a great deal, but many tug operators prefer to rely on feel.

Some tug operators refer to this feeling as being similar to the sensation of a water-skier at the end of a towline. The two prime sensory cues are the heeling angle and the sensation of digging into the water. This is similar to a water-skier shooting off to the side of a ski boat's wake. As the water-skier turns and increases the angle of the skis to the water flow, the ski edge digs in causing more hydrodynamic resistance and lift. This causes more pull on the towline. This creates a vertical heeling arm that, if left unchecked, pulls the skier over. The skier counters this pull with his body weight by leaning back. As long as the skier can keep these forces in balance he continues to accelerate off to the side. The more he moves off to the side, the greater these forces become.

In indirect towing, a tug operator may experience similar sensations. As he steers the tug to greater angles of water flow, the tug's resistance and subsequent pull on the tug's towing point increase. The vertical distance between the towing point and the center of hydrodynamic pressure creates a heeling arm. Heeling force is countered by the tug's inherent buoyancy, stability, and propulsion forces.

At some point, whether a tugboat or a water skier, the combination of hydrodynamic resistance and towline pull surpasses the capability of either to resist. The water-skier does not have enough body weight and physical strength, and the tug does not have enough stability and buoyancy, to counter the force of the heeling arm.

At this point both entities have three options. The first is to steer back toward the pulling force (ski boat or ship), which immediately reduces the angle to water flow and hydrodynamic resistance. The second is to reduce power and slow down. This immediately reduces the pulling force of the towline. The third is to release the towline altogether. If the skier and tugboat do not exercise one of these options, the skier may fall over and the tug may founder or capsize.

While these abort options may sound simple, they can prove difficult to execute. An operator can allow the tug to get beyond a recoverable position. The most recognizable sign of the tug's capability limit in indirect towing is the point of deck edge immersion. Once the deck edge begins to submerge in indirect towing, the designed stability and buoyancy calculations become invalid. At this point it is possible for the tug to be pinned in a vulnerable position. It lacks the power or maneuverability to counter the sudden exponential increase in heeling force. In these circumstances the tug may founder or capsize unless the towline is released.

It is important for both shiphandler and tug operator to appreciate fully how little time it takes to transition from being on the edge to over the edge of a tug's safe operating limit. While engaged in indirect towing it can be almost instantaneous. An accurate indication of towline force is essential to both the shiphandler for maneuvering and the tug handler for safety.

Jackknife Maneuver

Jackknifing is a maneuver allowing quick repositioning of the tug from indirect to direct towing. (Fig. 9–20). It is employed by tractor and ASD tugs. A typical application occurs when a tug is called to provide continuous steering forces as a ship slows down. As way comes off the ship the indirect towing position becomes more and more awkward. In order to continuously apply high towline forces the tug must come more in line with the towline causing it to lead sharply around the staple and perhaps begin to wrap around the deckhouse. At this point the tug operator can use a combination of towline tension, the render recovery feature of the winch and the tug's lateral maneuverability to swing around into the direct towing position. An experienced operator can execute this maneuver quickly with minimal loss of towline angle and

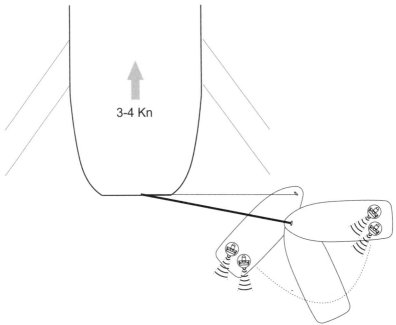

Fig. 9-20. ASD jackknife maneuver.

seamlessly transition to a direct towing position. Jackknifing is an appropriate maneuver at speeds of 3–4 knots.

When working a towline, the dynamic relationship between towing point, center of hydrodynamic pressure, and thrust creates a different feel for each tug design. The prudent tug operator will familiarize himself with his tug's controls, maneuverability, stability limits, and handling characteristics before engaging in towline shipwork. Handling the tug should be almost second nature.

The tug operator applying bollard pull to a towline is multitasking at a high level. He is responding to the pilot's orders, maneuvering the tug, communicating on the radio, monitoring the towline's strain, and calculating the best method to apply the amount and direction of bollard pull requested. All these responsibilities must be met while ensuring the safety of his personnel and his tug.

The safety of vessels and crew, as well as the pilot's reputation and the ship's welfare, often are in the tug operator's hands.

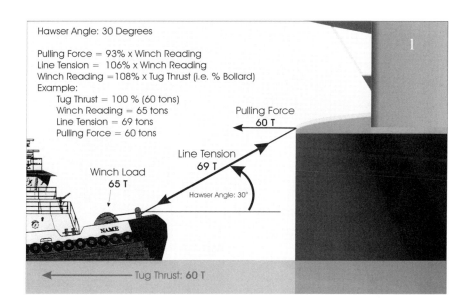

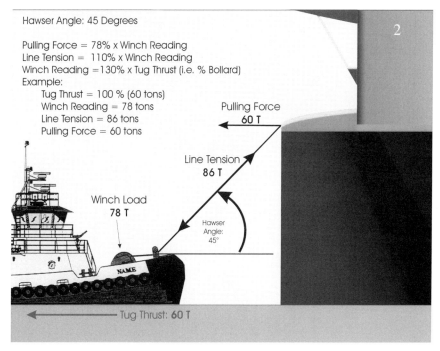

Fig. 9-21. Line loads on a short towline (image 1-4). *(Courtesy of Glosten & Associates, and Barry Griffin.)*

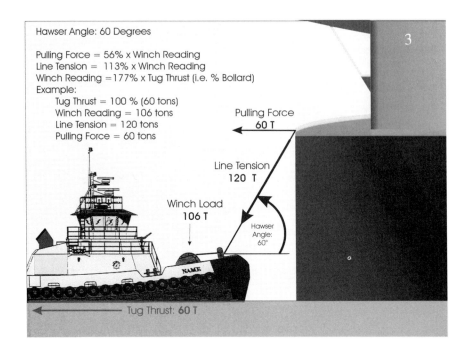

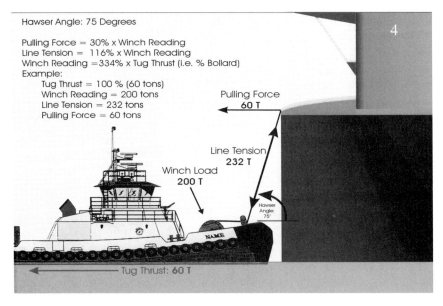

Fig. 9-21. *cont.* Line loads on a short towline. *(Courtesy of Glosten & Associates, and Barry Griffin.)*

Those hands should have an intimate feel for the tug, its capabilities, and limits on a towline.

SOME OF THE BEST PRACTICES OF THE EXPERIENCED TUG OPERATOR IN SHIPWORK

The following list is a summary of practices passed on by experienced tug operators. The good operator learns from his mistakes. The list below represents the positive outcome of the bent steel and bruised egos that accompanied these learning experiences. They are presented with the intent of helping tug operators pass from functional to ones with finesse.

1. Be wary of a line length that is too short. Some pilot's are tempted to ask for a short line length to hasten the tug's response time between pushing and pulling. Towlines or headlines with a steep angle put an undo strain on the line and its connecting gear (fig. 9-21). When the tug backs or pulls on the line it will be supporting the tug's weight and the bollard pull applied.

2. The tug operator should not back full on a slack headline or come ahead hard on a slack towline. The combined force of the tug's momentum and the power of its engine can part lines or pull marginal ship bitts and cleats out of the deck. The tug should come easily into its line (ahead or astern) until the line comes tight under the weight of the tug, at which point power can be increased gradually.

3. When pulling full on a line and an order to stop the engine is received, the tug operator should drop the engine to slow speed first and then stop. He must still be ready to check the tug's way should the elasticity of the working line propel the tug toward the ship too rapidly.

4. When docking a ship, the tug operator must never allow
    the tug to get caught between the dock and the ship. Stay in
    position as long as it is safely possible, notify the pilot of
    the situation, and remove the tug.
5. When a tug is coming alongside a ship that is underway,
    the tug captain should not allow himself to be distracted.
    The situation can change so quickly that even a moment's
    inattention can cause an accident.
6. Whenever a tug is going to work near the bow of a ship
    without putting up a line (especially at night), the tug
    operator should communicate via radio or whistle to let
    the mates on the bow know the tug's proximity. This may
    prevent the untimely dropping of the anchor on the tug.
7. When a tug crosses a ship's wake, the propeller has a
    tendency to cavitate. To avoid an accident, exercise caution
    when overtaking a ship in this circumstance.
8. When the after controls are to be used, the tug operator
    should ascertain beforehand that the covers are off and the
    controls are in neutral position. Fumbling with the control
    cover in the middle of the night can waste critical time.
    The controls should be in neutral position before shifting
    over from the wheelhouse. Some types of controls engage
    immediately (if they are in the ahead or astern position)
    when the control selector switch is changed. This has led to
    serious accidents.
9. Nylon-only lines should never be used for shipwork. They
    are much too elastic and can cause serious injury or death if
    they part.

COMMON HAZARDS OF SHIPWORK

Most marine casualties can be attributed to weather, mechanical
failure, or human error. The prime causes of accidents in shipwork
are mechanical failure and human error.

Mechanical Failure

Most mechanical failures can be attributed to a breakdown of the steering or the propulsion systems. If the engine fails to back or the steering does not respond, the ship or the tug is likely to be damaged. Lines and fenders also may fail, resulting in injury or damage. These failures may be unexpected but they are rarely acts of God. Many can be prevented by a comprehensive maintenance, inspection, and replacement program. Although the frequency and timing of mechanical failure may be unpredictable, the tug operator should be prepared with a mental contingency plan to guide his actions to minimize adverse effects.

Overall, the percentage of accidents attributable to mechanical breakdown is insignificant when compared to those due to human error.

Human Error

The human factor is the single, most frequent contributing factor in accidents in shipwork. Human misjudgment and poor decisions result in these common causes of accidents:

- excess speed
- poor handling (either ship or tug)
- careless line handling (ship or tug)
- faulty communication

Excessive Speed

When excessive ship speed is the cause of an accident, the tug often is the victim. It usually is no match for the ship in size or power. As mentioned before, tugs on towlines are easily overpowered, and can be dragged alongside the ship, damaged, or capsized. Ships moving too fast increase the risk of capsize for tugs using quarter lines. The hazard of stemming for tugs maneuvering around the bow also is exacerbated by increased ship speed. Even tugs working alongside a ship secured only by a headline have rolled over when the ship was operated or maneuvered at excessive speed. A shiphandler may ring for more speed if the ship making a bend is not responding well to its helm. This renders the tug completely ineffective. The ship may go aground, hit a dock, or strike another vessel.

Excessive tug speed also can be the cause of damage to both tug and ship. A tug coming alongside too fast may get caught in the vessel's wash (forward or aft) and take a dive toward the ship. A tug overtaking a ship may try to back down in the fast water of its wake, causing the propeller to cavitate with the same result.

Poor Ship or Tug Handling

In the case of faulty ship and tug handling, the ship often is the injured party. When a tug hits an inbound ship while coming alongside, haste on the part of the tug operator may be as much to blame as speed. In his rush to get alongside the ship, the tug operator may have lost focus on the task at hand or failed to allow enough time to feel the vessel out. Should a ship come ahead unexpectedly while a tug is working the stem with a line up, the tug easily could be caught and damaged under the flare of the ship's bow. Similarly, if a tug is made up dead astern, it can be caught under the aft overhang if the ship backs hard enough to pick up sternway. In these instances, the only thing the tug can do is cast off or cut its line.

Careless Line Handling

Quite a few accidents occur as a result of careless line handling by both tug and ship personnel. When the tug is let go, the ship crew should slack the lines back to the tug instead of letting go on the run. This prevents a line from striking the deckhand, falling in the water, or fouling the propellers of either vessel. Tug crews also should tend their lines carefully, especially the quarter lines and towlines. These lines can easily get caught in the tug's propulsion, particularly in the constantly turning propellers of controllable-pitch propellers.

Ships should stop their engines when casting off a tug's towline aft since it is easy for the ship's propeller to pick up such a line. This is especially true if the vessel has sternway.

Faulty Communication

Faulty communication can stem from poorly understood orders, radio failures, or radio transmissions garbled by overriding traffic. Voice transmissions from nearby ships can misdirect or confuse the tug's handler. Even whistle signals, echoing off of high buildings or other surrounding large structures, can be misinterpreted.

Shipwork can be a tricky business and the safety of all concerned depends on the awareness and vigilance of those who direct and handle the ships and tugs involved. The hazards discussed in this chapter are mentioned elsewhere in the text but the repetition is justified if it raises or renews the awareness of dangers associated with shipwork.

STUDY QUESTIONS

1. What are the two components of the Center of Hydrodynamic Pressure?
2. How should a tug's line be cast off from the ship to the tug?
3. Why is towline work on a conventional tug dangerous?
4. Explain how a conventional tug trips or girts.
5. How can the tug handler best avoid tripping or girting?
6. What is the advantage of using a tug alongside?
7. When does a conventional tug use a quarter line?
8. Explain a wrap line.
9. Can a tug work a wrap line on all vessels?
10. What is meant by the term snapping a tow?
11. Why should a tug made up for breasted towing have tight lines?
12. Why is indirect towing so powerful?
13. How is indirect towing used to advantage?
14. Why should a tug back easily on a slack headline?
15. What may happen when a tug crosses a vessel's wake?
16. How can cavitation endanger a tug crossing a ship's wake?
17. What are the principal causes of accidents involving tugs doing shipwork?
18. Of the causes cited which is the most common and why?

CHAPTER TEN

BASIC SHIPHANDLING PRINCIPLES

Much of the previous chapters focused on handling tugs and was presented from the tug operator's pilothouse view. This chapter marks a change in focus and perspective. We now leave the tug's wheelhouse and move up to the bridge of the ship. From this vantage point both tug operator and ship pilot can best view the application of tug forces within the context of shiphandling principles.

As noted, the tug can only push, pull, or apply passive resistance. It can apply these forces (subject to limitations) to the ship independently, in conjunction with, or against other forces influencing ship movement. The latter include wind, current, the ship's own momentum, propelling and steering forces. If these forces are applied intelligently, the tug can turn the vessel, assist it to steer, move it laterally, or check its way. In this instance the ship is like a lever with a fulcrum that shifts in accordance with the ship's motion and applied external forces. If a tug is used to turn the vessel, it is more effective when positioned farther away from the fulcrum. Conversely, if the tug is positioned closer to the ship's fulcrum it tends to move the vessel laterally.

Traditional shiphandling theory equates the fulcrum of this lever as the ship's pivot point. This serves as a useful tool to illustrate the principles of leverage in shiphandling. However, real life shiphandling does not neatly follow the illustrations in a book. The interrelation is dynamic between a ship's pivot point, turning fulcrum, leverage points and external forces in a fluid environment. It requires a seasoned shiphandlers eye to accurately comprehend the collective and changing effect of these factors and to use them to advantage in maneuvering ships.

Nonetheless, it is helpful to examine these shiphandling factors separately in order to assist in the development of the shiphandler's eye. Once understood, experience becomes the catalyst that can synthesize these singular understandings into a cohesive and accurate vision from the bridge of a ship.

Clear vision of a ship's behavior must include an understanding of three basic shiphandling principles:

- center of lateral resistance
- pivot point
- maneuvering lever

CENTER OF LATERAL RESISTANCE

The *Center of Lateral Resistance (CLR)* can be defined in engineering or practical terms. The shiphandler is more interested in the practical definition—the center of lateral resistance (CLR) is that point on a ship where the application of a lateral force results in the ship moving sideways with no rotation. When a lateral force (such as a tug pushing or pulling) is applied away from the CLR the CLR acts as the fulcrum of a lever that simultaneously rotates the ship and moves it sideways.

The center of lateral resistance can be visualized as a ball bearing set in bottom of the ship. The CLR rolls in accordance with the application of the ship's propulsion and external forces. When a lateral force is applied, the ship responds with varying amounts of lateral movement, dependent on the location of the sideways force in relation to the CLR. The degree of rotational force is dependent on the distance between the applied force and the CLR (fig. 10-1).

The center of lateral resistance location is not fixed and reflects the balance point of the following factors:

- hull shape
- direction of the ship's motion
- draft & trim

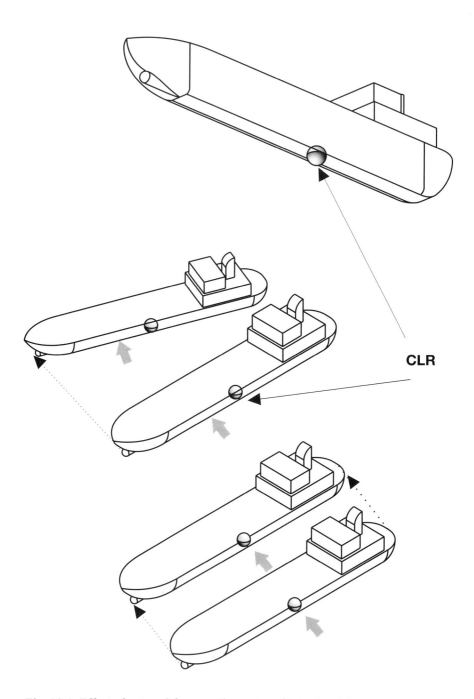

CLR

Fig. 10-1. Effect of external force on the center of lateral resistance.

The variety of shapes employed by naval architects to craft a ship's bow, mid-body, and stern generate varying degrees of lateral resistance depending on the direction of movement through the water.

When a ship is lying dead in the water with no motion it is only displacing, not resisting water. As soon as the ship acquires motion, either laterally, ahead, or astern, water begins to impede and resist this motion. In general, there is greater resistance at the leading end of the motion. As the ship gains headway the CLR moves forward. As the ship gains sternway the CLR moves aft.

A ship's draft has a direct bearing on the magnitude of resistance– as draft increases, hydrodynamic resistance increases. The same principle applies to the trim of the ship. More resistance is associated with the ship's end that has the greater surface area of submerged hull. The center of lateral resistance of a ship down by the stern is farther aft than one trimmed forward (fig.10-2).

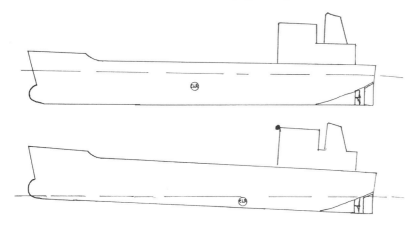

Fig. 10-2. Change in center of lateral resistance due to trim of vessel.

Of principal significance to the shiphandler is that the center of lateral resistance is a dominant factor in determining the allocation of rotational versus lateral motion that results from a lateral force applied to the ship. When lateral force is applied farther away from the CLR, rotation and lateral movement are produced; when lateral force is applied close to the CLR, minimal rotation and more lateral movement is the result.

The location of the center of lateral resistance may be, but is not often, in the same location as the ship's pivot point. Although the CLR has a direct effect on the location of the pivot point they are not often one and the same.

PIVOT POINT

The pivot point appears on the ship as an axis of rotation with reference to the water's surface. It is important to realize that pivot point location is the result of applied forces and the subsequent motion of the ship. It would be convenient to believe that there is always a one-to-one relationship between a particular ship motion (e.g. ship moving ahead) and an exact pivot point location (one-quarter ship length aft of the bow). But this is not the case. The location of a ship's pivot point migrates in accordance with the ship's motion, the pressure fields around the hull created by that motion, and the amount and location of forces applied to the ship. The pivot point shifts forward or aft toward the direction of the vessel's movement and away from the location of an applied lateral force (such as a tug pushing or pulling).

In general, the location of the pivot point tends toward the bow when the vessel is going ahead and moves aft toward the stern when the vessel gathers sternway. The reference for the ship's pivot point is the flowing water not the ground. When a vessel is not making way over the bottom but is stemming a current, the pivot point will still shift toward the up current end of the vessel.

Specific locations are associated with the following five basic ship movements (fig.10-3):

- stopped
- lateral
- headway
- sternway
- turning

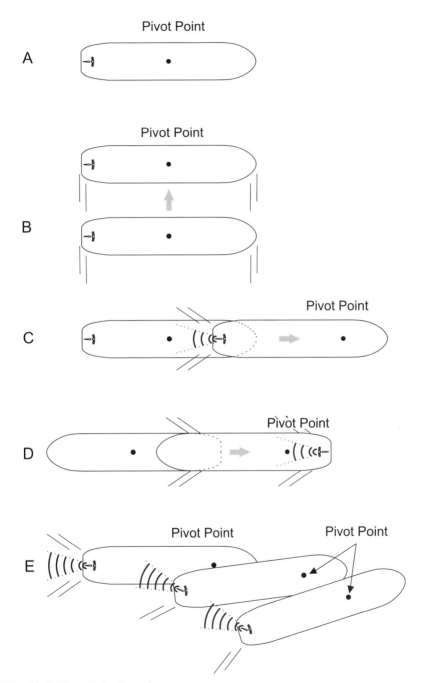

Fig. 10-3. Pivot Point Locations.

A ship that is stopped with no motion, has its pivot point located approximately amidships (fig. 10-3A). A ship with lateral and no rotational motion has its pivot point at the ship's center of lateral resistance (fig. 10-3B).

A ship that first starts to come ahead has its pivot point well forward. As the ship gains more headway, it generates more hydrodynamic resistance in the forward sections of the ship, causing the pivot point to settle in approximately one-quarter of a ship length aft of the bow (fig. 10-3C). If the ship slows and loses headway the pivot point begins to move aft, passing through amidships when the vessel is stopped. Once sternway is established, the pivot point moves to within one-quarter ship length forward of the stern (fig. 10-3D).

When a ship making way initiates a turn, lateral resistance begins to build on the ship's side. As the turn progresses, the ship simultaneously slides sideways and moves ahead, causing this resistance to build. The increased hydrodynamic pressure at the bow and along the ship's side moves the pivot point aft from its usual location associated with headway. It will lie at a point approximately one-third of a ship length aft of the bow (fig. 10-3E).

The pivot point is dynamic but it does not change location in large increments. It shifts as the balance between the ship's speed, direction, draft, and trim change as both internal and external steering and propulsion forces are applied to the ship.

MANEUVERING LEVER

A ship's maneuvering lever is defined by the relationship between two factors:

- pivot point location
- ship propulsion forces

Regardless of the origin of force, the efficiency of the maneuvering lever is a function of the distance between the applied force and the ship's pivot point. As described earlier, the amount of rotational force applied to this lever is a function of the distance between the applied force and the ship's center of lateral resistance. The maneuvering lever may either lengthen or shorten as the pivot point shifts in accordance with the ship's motion and the magnitude of the force being applied to the end of the lever. As ships are typically hundreds of feet in length, pivot point location has a tremendous influence on the effectiveness of the maneuvering lever.

The ship's propulsion forces have fixed locations. A primary source of propulsion is typically at the stern. Some may have additional auxiliary sources of propulsion such as a bow thruster.

In addition to the ship's own propulsion external forces may apply force to the ship. These additional sources may be environmental (the wind), mechanical (an anchor or mooring line), or dynamic (a tug). The shiphandler cannot control environmental factors and the use of anchors and mooring lines is limited to a few well defined situations.

However, the mobility and power of tugs offer the shiphandler a variety of options in applying external force to the ship. The art of shiphandling with tugs is the ability to apply the power of tugs within the context of the ship's pivot point and in conjunction with the ship's maneuvering lever. This is discussed in more detail in the following chapters.

STUDY QUESTIONS

1. What three fundamental principles of shiphandling are essential to a shiphandler's eye?
2. The Center of Lateral Resistance determines the resultant proportion of _____ vs. _____ force of an applied lateral force to the ship.
3. Describe pivot point location for a ship that is:
-   stopped
-   moving laterally
-   moving ahead
-   moving astern
-   turning
4. The pivot point functions as the _____ of the ship's maneuvering lever.

CHAPTER ELEVEN

STEERING AND PROPELLING SHIPS WITH TUGS

A primary function of tugs engaged in shipwork is to augment a ship's propulsion and steering. As previously noted, steering and propulsion are linked inextricably and the choice of tug position in ship assist should account for both factors. Once a ship has headway or sternway, most towing assist techniques will simultaneously affect a ship's direction and speed through the water.

A shiphandler's effective use of tugs to steer and propel ships requires an awareness and appreciation of three maneuvering factors. First, the pilot must understand the relationship between tug location and the ship's pivot point. Second, he must be cognizant of the orientation of both the tug's position and the ship's propulsion point in reference to the ship's pivot point. Third, he must account for the strengths and weaknesses associated with the four basic tug positions in ship assist.

TUG TO SHIP PIVOT POINT DISTANCE

When the tug applies a steering force it creates an assist lever (AL) between the tug and the ship's pivot point. Positioning the tug at the greatest possible distance from the pivot point can maximize this effect. In general, positioning a tug at or near the stern is the optimal position to steer a ship with headway. Positioning a tug at or near the bow is most effective steering a ship with sternway (fig. 11-1).

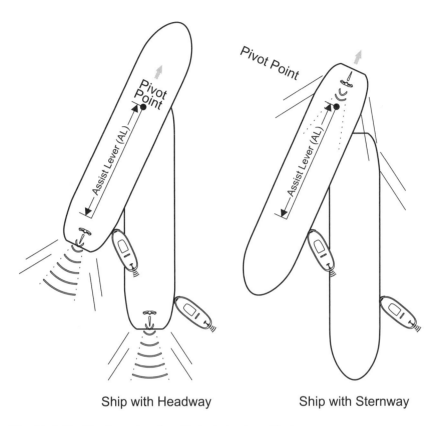

Ship with Headway   Ship with Sternway

Fig. 11-1. Positioning a tug for efficient steering effect.

RELATION BETWEEN TUG, SHIP PROPULSION POINT, AND PIVOT POINT

Ship pilots use assist tugs to best advantage by using the tug's AL to enhance, oppose, or lie neutral in relation to the ship's maneuvering lever (ML). The ship pilot must visualize the two levers and their orientation to each other. Ship speed, direction, and tug location determine the length of the two levers and whether they lie opposite or on the same side of the ship's pivot point. The pilot can manipulate the two levers through helm, engine, and tug orders to make the ship turn, stop, or move laterally (fig. 11-2).

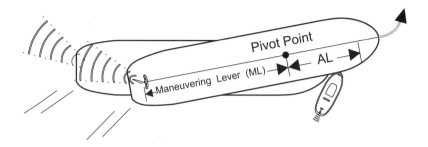

Fig. 11-2. Tug assist lever and ship maneuvering lever.

TUG POSITIONS

Chapter 9 discussed the four basic tug assist positions:

- leading tug towing on a line
- tug alongside the ship pulling on a line or pushing
- tug alongside the ship breasted or "on the hip" towing
- trailing tug, pulling on a line or pushing

Each position has its own advantages and disadvantages when applying steering and propulsion forces to a ship. For the sake of clarity, the discussion of these positions assume only one tug is assisting the ship. Balancing the force of multiple tugs is discussed in chapter 15.

LEADING TUG TOWING ON A LINE

A lead tug can apply considerable steering force to a line. It can apply equal force to port or starboard and is unrestricted by the ship's flair or tumble. The tug's response time to an order for a change in direction is not immediate. It is related to the tug's ability to shift laterally to the new position that creates the desired towline angle. The tug must remain forward of the ship, at any but the lowest ship speeds, to keep pace with the ship's advance. A tug in this position always has a force component that adds to ship speed, regardless of towline angle.

Lead Tug—Ships with Headway

The pivot point of a ship with headway lies approximately one-quarter of a ship length aft of the bow. The assist lever of a tug on a towline is considerably shorter than the ship's maneuvering lever (fig. 11-3).

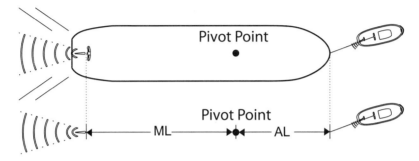

Fig. 11-3. Assist lever and maneuvering lever of tug towing a ship with headway.

The ship, on the other hand, has greater leverage and more power. This is why it is so easy for a ship making headway to overwhelm a tug on a towline. However, with prudent use of the ship's rudder and propulsion, the ship pilot can use a tug on a towline to control the ship's bow and regulate the turn rate of the ship. This is an effective configuration for controlling both ends of the ship. The ship's propulsion controls the stern, and the tug assists in control of the bow.

Using the tug's assist lever and the ship's maneuvering lever in tandem can increase the ship's turn rate (fig. 11-4A). The tug's assist lever can also be used to counter the ship's maneuvering lever. This can assist in slowing or checking the turn of the ship (fig. 11-4B).

Lead Tug—Ships with Sternway

The pivot point on a ship with sternway lies approximately one-quarter ship length forward of the stern. Due to its proximity to the ship's propulsion point, this creates short or negligible levers for both the ship and a lead tug on a towline (fig. 11-5). The reduced effectiveness of a ship's rudder going astern, coupled with the torque of its propeller, also diminish the ship's maneuverability.

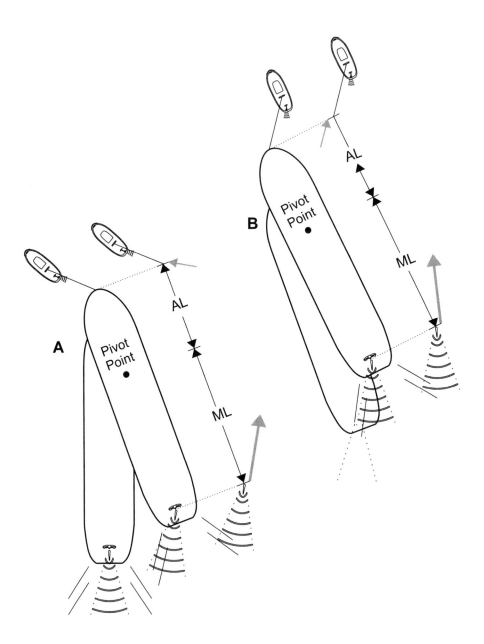

Fig. 11-4. Using the assist lever to increase or check the ship's rate of turn.

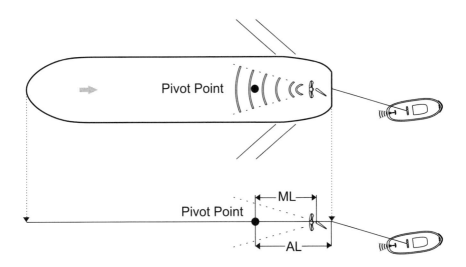

Fig. 11-5. AL and ML lead tug towing a ship with sternway.

Additionally, wind and other external forces have tremendous leverage on the bow of the ship.

The factors noted in the preceding paragraph make it difficult if, not impossible, for one tug on a towline to effectively steer the ship with sternway. The tug has neither the leverage nor the power to counter the ship sheering. At best, the tug provides a propelling force that allows the ship to move astern with minimal speed. The ship pilot must steer the ship by judiciously using the rudder and small kicks ahead. These kicks ahead should be communicated to the tug operator in advance so he can anticipate the effect of the ship's propeller wash on the tug.

Certainly this is not an ideal arrangement for steering a ship with sternway. A more prudent arrangement is to have two tugs: one towing on the ship's stern and one near the ship's bow. The lead tug provides propelling force to advance the ship slowly astern. The bow tug uses its long assist lever to apply steering forces. And the ship uses its rudder to shift the ship's stern (fig. 11-6).

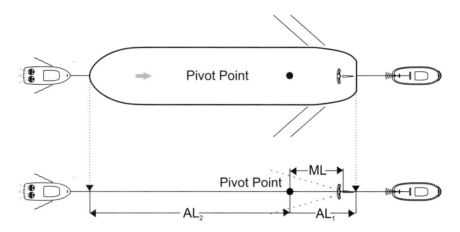

Fig. 11-6. Two tugs assisting a ship with sternway.

TUG ALONGSIDE THE SHIP PULLING ON A LINE OR PUSHING

Tugs working alongside ships are used principally to help steer and move the ship laterally. Any assist in ship propulsion usually is a secondary effect of its steering action.

As described in chapter 9, the tug working alongside may put up one, two, or multiple lines to assist in holding its position perpendicular to the ship's side. This is appropriate when the ship is at slow speeds and being breasted to or from a berth. In steering situations speed may be higher, and it is vital that the tug be able to lay as flat as possible alongside the ship. Headline length is critical and must be long enough to allow the tug to lay parallel to the ship's side (fig. 11-7). If line length is too short the tug will be snubbed in at an angle and the line may part as ship speed increases or the tug backs.

A tug working alongside a ship has three options of how to apply force: it can push, act as a drogue, or actively pull on its line.

Pushing

When pushing, the tug swings out as close as possible to a 90 degree angle to the ship's centerline. Unless the tug is at this angle there is

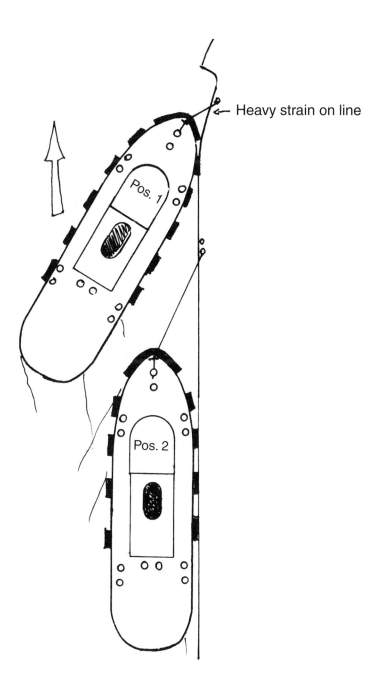

Heavy strain on line

Fig. 11-7. Proper headline length for alongside.

a forward as well as lateral force component to the tug's push. This adds to the ship's rate of advance. Once in position the tug can apply full power without concern of parting lines or pulling cleats and bitts up from the ship's deck.

Drogue

The tug alongside can act as a drogue. When a tug lays passively on its line with engines stopped, the line is under tension created by the tug's weight and hydrodynamic resistance. The tug's drogue effect has a tendency to turn the ship toward the tug side of the ship. If this is an unwanted effect the tug can come ahead, ease the line tension, and run with a slack line.

Pulling

The tug can apply force by pulling on its line. As in pushing, the most effective position for creating a lateral force is when the tug maintains a 90 degree line angle to the ship. At lesser angles the lateral force is gradually replaced by a retarding force which detracts from the ship's rate of advance.

Pulling vs. Pushing

A tug alongside generally is not as effective pulling on its line as it is pushing into the ship. This difference depends on how well a tug maneuvers astern and its loss of efficiency when operating in reverse. Maneuverability astern is a key to the tug getting close to the optimal 90 degree angle. The tractor and ASD tugs have the advantage here. However, all tugs alongside lose some efficiency pulling on a line. This is due to the tug's propulsion wash hitting both the underbody of the tug and the side of the ship's hull.

Ship Speed

Tug performance alongside depends on the ship pilot's prudent control of ship speed. A two-knot increase in ship speed can make the difference between the tug being effective or serving as no more than an expensive drogue. Excessive ship speed has a cascading effect. It dictates the amount of drag the tug must overcome to swing into position. It also determines how close a tug can get to the optimal 90 degree angle and limits the amount of power the tug can apply to its line before it parts.

Response Time

A tug alongside is quicker than a lead tug on a towline when responding to the ship pilot's orders. When called to back, the time for the tug to get on the line is largely a function of line length. The longer the line the more distance the tug must cover before the slack is taken out.

Positioning a tug for work alongside may be limited by the ship's overhang and chock configuration. It may not be possible to place the tug in a position that creates both the desired distance between the tug and the ship's pivot point and propulsion point.

Working tugs alongside can be very effective in providing steering assistance to a ship. However, the ship pilot must be aware of the strengths and weaknesses associated with the tug's specific location and action alongside the ship.

SHIP WITH HEADWAY

Tug Alongside Bow Shoulder

A tug alongside near the forward shoulder of a ship with headway is at, or almost, abeam of the ship's pivot point. This position gives the tug a small to negligible assist lever to turn the ship. However, a tug in this position can steady and apply lateral forces to the ship's pivot point. When pulling on its line, the tug can function as a dynamic springline and enhance the effectiveness of the ship's own steering and propulsion.

Pushing

When the tug is on the outboard side of a turn it can push to assist the ship's maneuvering lever. As the ship rotates around its pivot point the tug pushes the ship's bow towards the turn. At the same time the ship's propulsion drives the ship's stern away from direction of turn (fig. 11-8). Initially the tug has a positive turning effect on the ship.

However, the tug's effectiveness in this configuration can be deceptive. Although it may appear that the tug is applying a force parallel to its centerline, the actual effect on the ship may be entirely different. This is due to the dynamic balance between four factors:

- angle of the tug
- offset position of tug
- drag force
- tug's applied horsepower

When the tug pushes on the outboard side of a turn, three of these factors enhance, and one detracts, from the ship's turn rate (fig. 11-8).

The easiest factor to visualize is the tug's angle to ship's centerline. The tug's angle represents the resultant vector of two components: forward as well as lateral propelling forces. However, even if the tug is parallel to the ship's side pushing on its headline, a lateral component still remains, forcing the ship away from the tug.

This is due to the rotational lever created by the tug being laterally offset from the ship's pivot point. When the tug pushes on the outboard end of this lever, it tends to rotate the ship away from the tug. These two factors, angle of the tug and offset position, assist in the turn.

As the tug assumes a greater angle to the ship, the tug's hull creates a drag force. This force acts similar to the drogue effect of the tug laying alongside. In the case of the tug on the outboard side of a turn, this drag force detracts from the ship's turn rate.

Much of the balance between the above factors is determined by the amount of horsepower applied by the tug. An increase or decrease in the tug's engine rpms may strengthen or weaken the effect of the other factors.

The most common effects of a tug pushing on the forward outboard side of a ship are an increase of the ship's turn rate and speed of advance. However, in some situations even with the tug pushing full at 90 degree, the ship turns toward the tug rather than away. These are circumstances in which hydrodynamic forces associated with the ship's turn work through the ship's maneuvering lever to override the tug's assist force.

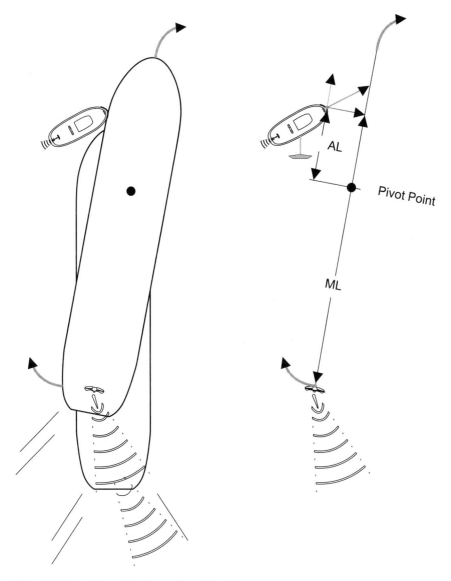

Fig. 11-8.Tug alongside forward-pushing into a turn.

Backing

A tug on the forward inboard side of a turn can assist by backing on its headline (fig. 11-9). In this configuration all four factors assist in the ship's turn rate. The angle of the tug, its force on the assist lever, its drogue effect, and its applied bollard pull cant the bow toward the inboard side of the turn. These forces all enhance the effectiveness of the ship's propulsion.

The tug backing on its line slows the ship. At slower speeds the ship's propeller and rudder have more of a steering, than a propelling, effect. Additionally, the tug backing on its line acts as a dynamic springline. This tends to increase the effectiveness of the ship's maneuvering lever (fig. 11-9).

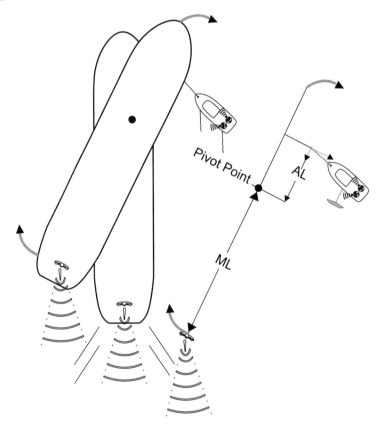

Fig. 11-9. Tug alongside forward-backing into a turn.

The weak link in this configuration is the tug's line. The shiphandler should be mindful of ship speed when ordering a tug to back on a line. If the ship is traveling too fast, the weight of the tug being towed alongside may impose such a severe strain on the line that it might part if the tug backs hard.

Tug Alongside Stern Quarter

Positioning the tug alongside the aft quarter has its advantages and disadvantages. The principal advantage is that the tug is a good distance away from the ship's pivot point and has a long assist lever. The disadvantage is that the tug cannot apply equal force toward and away from the ship.

Both the tug's assist lever and the ship's maneuvering lever lie on the same side of the pivot point. When the two work together the tug accentuates the effect of the ship's steering and propulsion. The best results are attained when the tug is on the inboard side of the turn and pushes on the ship (fig. 11-10). This intensifies the ship's rudder effect, moving the ship's stern away from the direction of the turn. Because the tug is on the inside aspect of the turn, the tug's hull resistance also adds to the turning effect.

A secondary but less effective means of assisting in this position is for the tug to lay flat alongside the ship and back on its line. This assists in the turn by twin-screwing the ship (fig. 11-11). In this position the tug's hull is dragged sideways through the water by the ship's swinging stern. This extra hydrodynamic resistance detracts from the turn rate.

A conventional tug positioned on the aft quarter on the outboard side of a turn is practically useless (fig. 11-12). All four factors oppose the turn. Unless the ship has bare steerage way the conventional tug will oppose the turn even when backing full. At steering speeds the conventional tug allocates most of its horsepower to maintaining position. There is minimal lateral pull applied to the tug's headline. Instead of enhancing the turn, the conventional tug's angle, offset position, drag, and horsepower are all working on a long assist lever to counter the turn. This unwanted effect has surprised many shiphandlers who are not experienced in shiphandling with tugs.

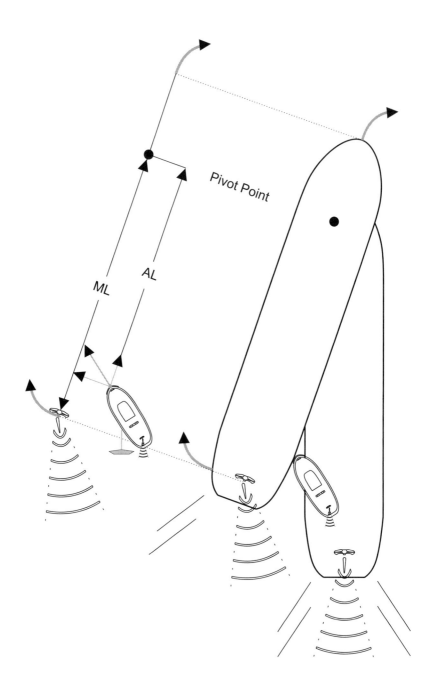

Fig. 11-10. Tug alongside aft quarter pushing.

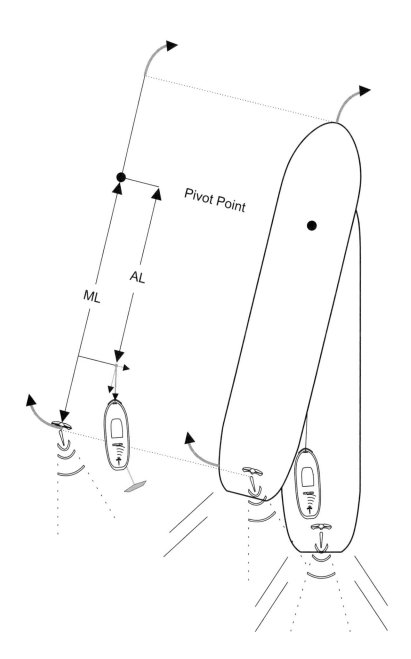

Fig. 11-11. Tug alongside aft quarter backing inboard side of turn.

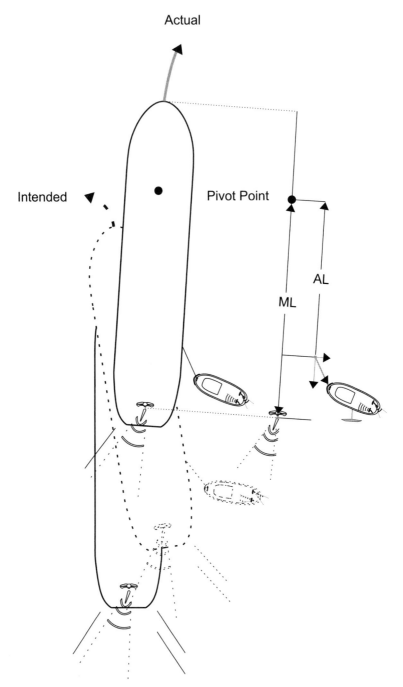

Fig. 11-12. Conventional tug alongside aft quarter backing outboard side of turn.

Tractor and ASD tugs alongside the ship's quarter have an alternate option. As long as ship speed remains at the tug's direct towing capability, the tractor and ASD tugs can position themselves alongside the ship, propulsion end first (fig. 11-13). This gives these tugs two options to assist the ship. If they are on the inboard side of a turn and required to push, they can direct their thrust toward the ship while laying alongside. If they are on the outboard side of a turn, they can direct pull at about a 45 degree angle to the ship. This direction of line pull both advances the ship and rotates the ship's stern around the pivot point.

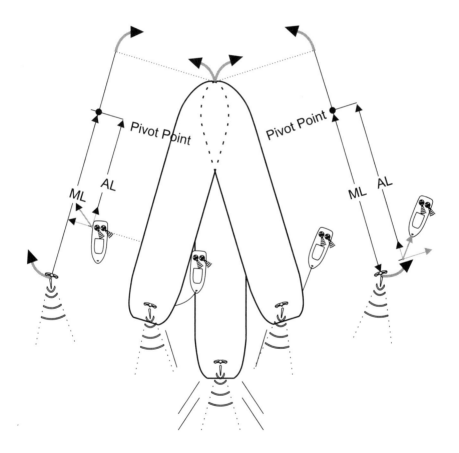

Fig. 11-13. Tractor/ASD alongside aft quarter pulling outboard side of turn, pushing inboard side the turn.

In this configuration the tractor and ASD tugs are maneuvering away from the ship's counter. This may offer the additional advantage of being able to use a ship chock located father aft than one accessible to a conventional tug.

SHIPS WITH STERNWAY

Tugs alongside Ship's Bow Shoulder

When the tug is alongside the bow shoulder of a ship with sternway, the tug is a good distance from both the ship's pivot point and propulsion point. The tug lies on one side of the pivot point while the ship's propulsion lies on the other. This creates a short maneuvering lever for the ship and a long assist lever for the tug.

The tug can apply this leverage effectively by pushing on the ship (fig. 11-14). Pushing on the ship's port bow swings the bow to starboard and assists in turning the ship to port. The pilot also can steer the ship by using the ship's rudder and propeller to give kicks ahead with the desired rudder angle and use the tug to steady the bow. The conventional tug can work most effectively against the ship's steering and propulsion by pushing, while the tractor or ASD tug can push or pull. The conventional tug backing in this position has the same limited, or contrary, effect as it does alongside the stern quarter of a ship with headway.

Tug Alongside Ship's Stern Quarter

With the tug alongside the stern quarter of a ship with sternway, the tug is close to both the ship's pivot point and propulsion point. The tug has an ineffective assist lever but can assist in moving the ship's pivot point laterally. In this position the tug works by pushing or pulling in conjunction with the ship's rudder to guide the ship's stern (fig. 11-15).

This arrangement provides poor, if any, control of the ship's bow. It is rarely used unless the ship has a bow thruster or there is an additional assist tug at the bow.

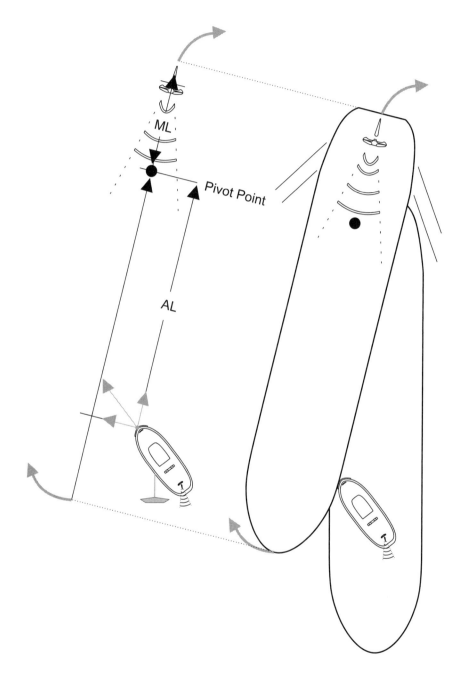

Fig. 11-14. Tug alongside bow, ship with sternway.

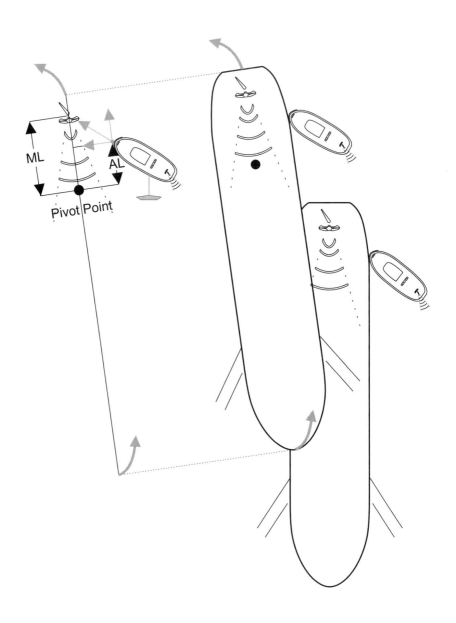

ML

AL

Pivot Point

Fig. 11-15. Tug alongside stern quarter, ship with sternway.

TUG ON THE HIP

On the hip or hipped up are terms commonly used to indicate that the tug is solidly lashed to a vessel. The correct nautical terminology for this practice is breasted or alongside towing.

Although this practice primarily is used in shifting barges, it also is useful in shipwork. This position differs from the previously discussed alongside positions in that the tug, by the use of its lines, becomes one with the ship. In this position the tug can assist in both steering and propelling the ship.

Although the tug is solidly connected to the ship, its position is offset from the ship's centerline. This alters the dynamics of the tug's application of steering and propelling forces. The tug's offset position creates a rotational lever. When the tug comes ahead or astern, it creates both a propelling force and a rotational force. If the tug comes ahead, with its helm amidships, the ship will gather headway and turn away from the tug (fig. 11-16A). If the tug backs, the ship will lose headway and come toward the tug (fig. 11-16B).

The effect of the tug's off-center position can be countered by the steering of either the tug or ship. When the tug alongside is the sole provider of a ship's steering and propulsion the ship tends to skid, or set to the side more than a ship operating under its own power (fig. 11-16C). This is caused by the large angle of rudder required for the tug to keep the ship under control and to compensate for the tug's off-center location. This is more apparent in light draft vessels than in those that are deeply loaded.

TUG MADE UP ALONGSIDE SHIP'S STERN

The effect of lashing the tug alongside a ship's stern creates a twin-screw ship with independent rudders. The tug's assist lever nearly matches the length and orientation of the ship's maneuvering lever. Consequently, the tug shares the same advantages and disadvantages as the ship's propulsion when operating with either headway or sternway. However, the ship, in conjunction with the tug, can be twin-screwed one way or the other to steer the ship. A tug on the

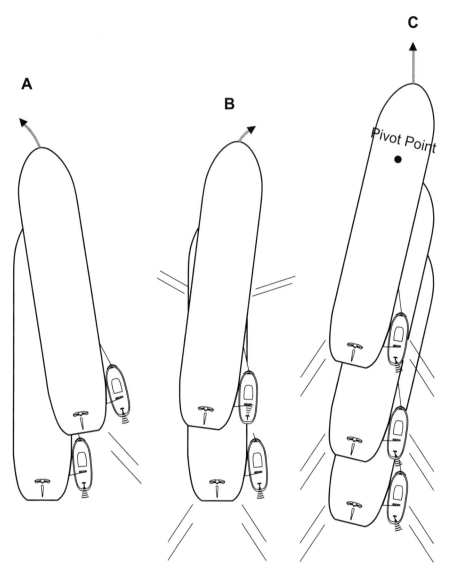

Fig. 11-16. Effect of tug in breasted towing. A: Tug comes ahead. B: Tug backs down. C: "Skidding"

inboard side of a turn backs its engine, while the ship comes ahead with its helm turned toward the tug. A tug on the outboard side of the turn comes ahead with its helm toward the ship while the ship backs its engine (fig. 11-17).

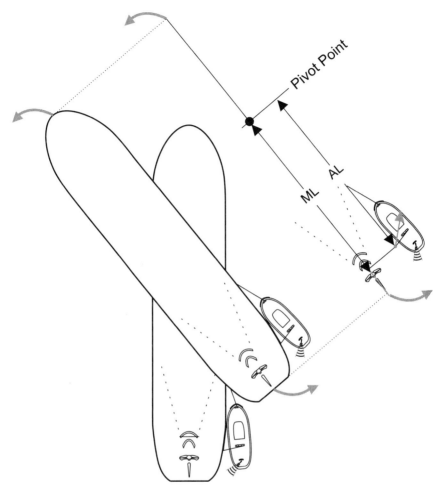

Fig. 11-17. Twin-screw effect, breasted towing alongside ship stern.

TUG MADE UP ALONGSIDE SHIP'S BOW

The tug functions as a steerable bow thruster when made up alongside the ship's bow. When the ship has headway, the tug is close to the ship's pivot point and assists by working its thrust to steady or move the pivot point laterally. Unless the tug has omni-directional thrust capabilities, this configuration is effective only at slow speeds.

When the ship has sternway, the tug has a long assist lever and steers the ship by swinging the ship's bow to port or starboard (fig.

11-18). The tug also can be used to work against the ship's steering and propulsion when the pilot gives orders for corrective kicks ahead with the ship's rudder and propeller. A particular advantage of this configuration is that the tug can function as the prime propelling force for the ship. It can advance the ship at a much slower speed than if the ship backs under its own power. The lower speed increases the effectiveness of both the tug's steering forces and the ship's kicks ahead, with the helm over.

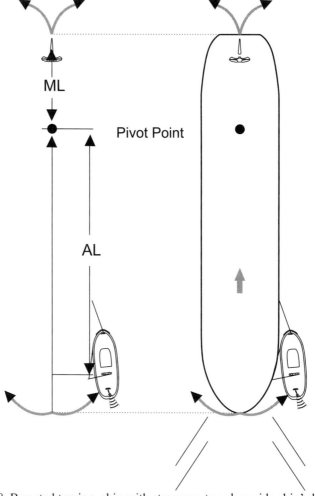

Fig. 11-18. Breasted towing, ship with sternway, tug alongside ship's bow.

TUG DESIGN FACTORS IN BREASTED TOWING

When the tug is made up alongside, its performance limits are dictated by the tug's type of steering and propulsion. This is an important consideration when choosing where to position a specific type of tug. As noted before, the conventional tug is most effective when it propels and steers by coming ahead on its engines. It is strongest in breasted towing when made up alongside with its bow facing the direction of ship advance. However when it is made up with its stern facing the direction of the ships way, two limiting factors come into play. The first is its limited bollard pull and maneuverability astern. The second is the effect of negative water flow.

The tug's propulsion operates in positive flow when the propeller wash is moving in the same direction as the water flow created by the hull movement. When the tug's wash opposes the general direction of the water flow, it is operating in negative flow conditions.

Negative flow considerably increases the torque loadings on the propeller and engine. The conventional tug made up alongside the bow of a ship with headway is operating in negative flow conditions when the tug comes ahead. At higher speeds this increased load may be enough to stall, or damage, the tug's engine, clutch, or couplings. The ship pilot must keep ship speed to a minimum in such circumstances.

The omni-directional thrust capabilities of the tractor and ASD tugs give the tug operator much more flexibility in managing negative flow conditions. They also compensate for the effect of the tug's offset position from the ship's centerline.

TRAILING TUG, PULLING ON A LINE, OR PUSHING

A trailing tug is in the best position to steer the ship. This position creates the longest assist lever and allows the tug to apply equal forces to both starboard and port. The tug can enhance the effect of the ship's rudder on a ship with headway and can also control the bow and steer a ship with sternway.

This position offers a long and effective assist lever. However, the tug's performance capability at the end of this lever directly correlates with the location of the tug's towing point and type of propulsion.

The conventional tug can provide steering assistance as a trailing tug by using either its towing point aft or its bullnose forward. Towline work as a trailing tug is limited by the ship's speed. At slow speeds (one to three knots) the tug can lie broadside to the general direction of the ship's advance (fig. 11-19A). Single-screw tugs require the use of a gob line to stay in position. Twin-screw tugs may be able to hold position with engine and rudder alone. The tug can back or come ahead to apply a steering force to the ship. At slightly higher

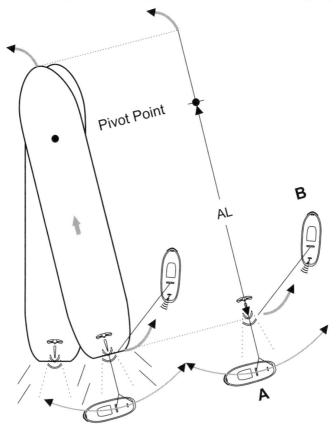

Fig. 11-19. Trailing tug. Conventional tug on a towline.

speeds the conventional tug can offer steering assistance to one side only. The tug must face the direction of the ship's advance and tow in a position where its towline leads forward from the ship. This adds both a propelling and turning force to the ship (fig. 11-19B).

When a conventional tug is operating stern first on a towline, the ship pilot must carefully manage ship speed and propulsion wash. In these arrangements, the tug is at a most vulnerable position at all but the lowest ship speeds. At speeds above one to two knots the tug, as pictured in fig. 11-19B, cannot safely transition from one side of the ship to the other, or to the broadside position in fig. 11-19A. The tug's broadside aspect of its towline pull can easily heel and possibly capsize the tug at excessive ship speeds. In addition, when the tug is trailing the ship's stern, the tug is directly exposed to the ship's propeller wash. This turbulent, increased water flow of the wash also can create too much towline force and subsequent heeling of the tug.

A conventional tug also can function as a trailing tug by working off its bow. A single, split, or wrapped headline enables the tug to apply force quickly and equally to both port and starboard. As a trailing tug on a ship with sternway, the conventional tug pushing on the ship's stem with a wrapped headline can both steer the ship and offer resistance when the ship's engine kicks ahead.

The tractor and ASD tugs are well suited for the role of the trailing tug. Unlike the conventional tug, these tugs can provide effective steering forces at higher ship speeds. Depending on the specific design, a tractor or ASD tug can use direct towing at ship speeds of five knots or less, and indirect towing at higher speeds (fig. 11-20). These techniques apply both a steering and retarding force to the ship.

At its essence, employing tugs to steer and propel ships is a matter of creating and using the tug force as an effective lever in conjunction with the ship's maneuvering lever. Understanding this principle of leverage can make the difference between the tug facilitating the ship's intended heading or course changes, versus unintentionally opposing them. The visualization of these levers creates the mental context to position and use tugs in the most effective manner.

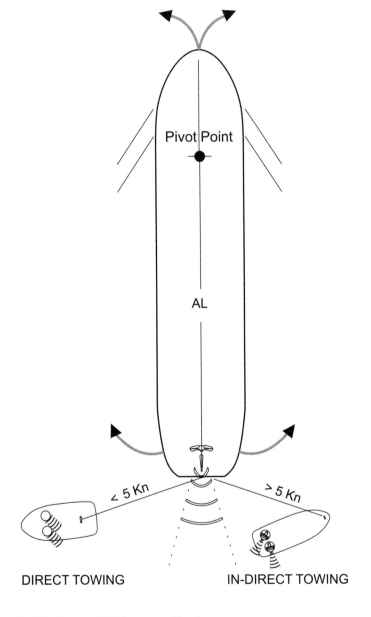

DIRECT TOWING   IN-DIRECT TOWING

Fig. 11-20. Tractor/ASD tug as trailing tug.

STUDY QUESTIONS

1. Define a tug's assist lever.
2. Why is it easy for a ship with headway to overwhelm a lead tug on a towline?
3. What are the three means a tug has to apply force when working alongside?
4. What are the four factors that affect a tug's effectiveness when working alongside?
5. Why is a conventional tug backing on the stern quarter, on the outboard side of a turn, ineffective in turning the ship?
6. In breasted towing, a tug made up on the ship's stern creates, in effect, a _____ ship.
7. In breasted towing, a tug made up on the bow functions as a _____.
8. What is negative water flow?
9. How is negative water flow a factor in conventional tug performance?
10. Why is a trailing tug the most effective position for steering a ship?
11. What precautions should the shiphandler exercise when employing a conventional tug as a trailing tug on a towline?
12. What two methods can tractor and ASD tugs use to steer a ship as a trailing tug?
13. What is the essence of effectively using tugs to steer and propel ships?

CHAPTER TWELVE

TURNING AND LATERALLY
MOVING SHIPS WITH TUGS

The principles of leverage used in positioning tugs to assist steering of ships apply equally to turning ships and moving them laterally. The length of the ship represents a lever, while its pivot point represents the fulcrum. The ship pilot turns the ship or moves it laterally by manipulating the forces applied to the fulcrum or to one or both ends of the turning lever. Proper placement of tugs in this context, and judicious use of the ship's steering and propulsion, produce the desired movement of the ship.

EFFECT OF TUG ON SHIP'S PIVOT POINT

The interrelation between the ship's center of lateral resistance (CLR) and its pivot point is most apparent when the tug pushes or pulls laterally on a stopped ship. The force applied by the tug has much more influence on the pivot point of a stopped ship than a ship that is making way. The pivot point may coincide with the CLR when the ship is stopped, but once a lateral force is applied the pivot point moves in response to the applied force. This makes the pivot point a somewhat moving target. A good example of this is one tug pushing on the end of a stopped ship. The ship's pivot point moves from midships (over the CLR) to the ship end opposite of the tug (fig. 12-1). This creates a long assist lever for the tug in which the ship's pivot point is the fulcrum, not its center of lateral resistance.

One tug pushing on the center of lateral resistance of a stopped ship keeps the pivot point on the CLR of the ship. This produces lateral motion of the ship. When multiple tugs push or pull in combination with the ship's steering and propulsion, the dynamic balance between all these forces determines pivot point location. Tracking the changes in pivot point location is crucial to the ship pilot in order to use the leverage of tug and ship to advantage.

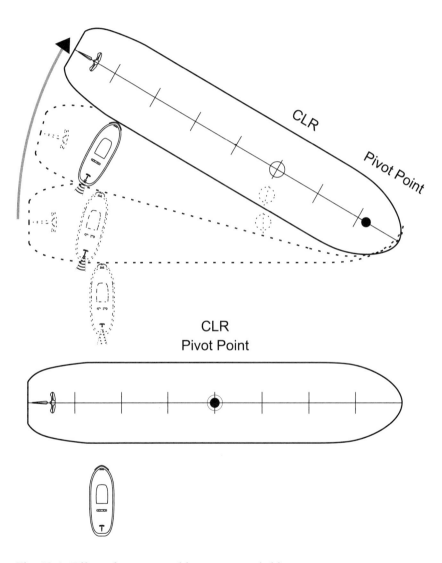

Fig. 12-1. Effect of one tug pushing on stopped ship.

TURNING STOPPED SHIPS WITH TUGS

Using tugs to turn ships is a matter of pushing or pulling on one or both ends of a turning lever. This turning lever is determined by the ship's pivot point. The greater the tug's distance from the ship's pivot point, the greater the tug's leverage.

A force applied to one end of the turning lever turns the ship in approximately two ship lengths. When a tug pushes on the bow of a stopped ship, the ship will rotate around the pivot point near the ship's stern. As the ship gains turning momentum it also begins to creep ahead. This forward movement can be countered by the ship backing as needed, limiting the diameter of the turning circle to approximately two ship lengths (fig. 12-2).

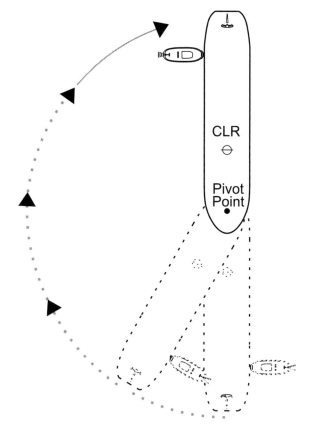

Fig. 12-2. Turning, stopped ship with one assist tug.

Forces applied to both ends of the turning lever can turn the ship within one ship length. Dual forces, working on opposite ends of the lever, but in the same rotational direction, rotate the ship around its center of lateral resistance which, in this case, functions as the pivot point. The applied forces can consist of two tugs, or one tug working in conjunction with the ship's steering and propulsion. Either way both forces must be in balance to produce the desired turn (fig. 12-3).

If equal forces are applied to both ends of the ship, the ship pivot point remains in the center of the ship. If one force is stronger, the pivot point shifts toward the weaker force. These complementary forces must match each other in sideways and fore and aft thrust.

When the ship's steering and propulsion are used at one end of the turning lever they create a forward and turning motion in the ship. The opposing tug must counter this forward motion by adjusting the angle of its towline pull or push (fig. 12-4).

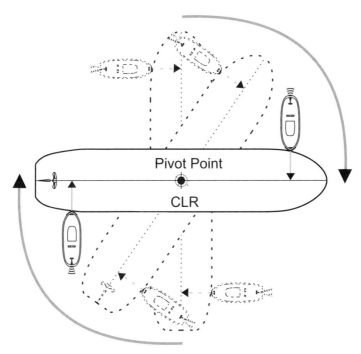

Fig. 12-3. Turning, stopped ship with two assist tugs.

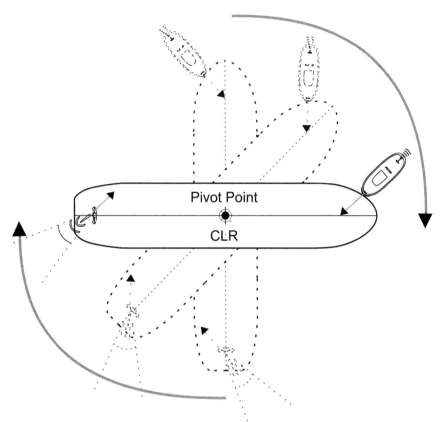

Fig. 12-4. Turning, stopped ship with one assist tug and ship's propulsion.

LATERALLY MOVING STOPPED SHIPS WITH TUGS

Moving a ship laterally requires the absence of rotational consequences of forces applied to the ship's turning lever. This can be accomplished in two ways: by applying force directly to the fulcrum or both ends of the turning lever.

One tug pushing abeam of the center of lateral resistance on a stopped ship is directing its force at the ship's pivot point. Pushing on the turning lever's fulcrum produces lateral motion of the ship (fig. 12-5).

Two equal, yet opposing forces, applied at each end of the ship's turning lever, produce the same effect. Each negates the other's rotational force on the lever, producing lateral movement of the ship. The two opposing forces can be either two tugs (fig. 12-6) or one tug working against the ship's propulsion (fig. 12-7).

The above examples have focused primarily on simplified situations, in which the ship has no way on. Although a ship pilot does encounter such situations, much of his application of the art of shiphandling takes place in the ship having way. He can still turn and move the ship laterally, but in this context he also must account for the effect of the ship's motion as well as the forces applied by the tugs. The ship's pilot should be familiar with the following three principles of ship movement:

- pivot point movement
- drift angle
- lateral motion

Pivot Point Movement

The pivot point of the ship migrates to different positions in sequence with the phase of the ship's turn (fig. 12-8). A ship lying dead in the water, with no wind or current, has its pivot point located at its center of lateral resistance, approximately midships. If the pilot turns the rudder hard over and orders the ship's engines ahead, he applies a force that both turns the ship and moves it ahead. The ship's stationary inertia immediately resists the ship's forward motion, causing the pivot point to move initially within one-eighth ship length of the bow. In these first moments, the ship's rudder is working on a long lever and turns the ship with minimal forward advance. This is a moment all pilots treasure: turning a vessel under its own power without using the limited space that may lie ahead. The moment can sometimes be short-lived. As soon as the ship begins to gain headway, water resistance builds at the ship's bow. This forces the pivot point aft to a position about one-quarter ship length aft of the bow. This migration shortens the ship's turning lever, reducing the rudder's turning effect. As the ship continues in the turn and water resistance spreads to the ship's side, the pivot point moves to within one-third ship length of the bow.

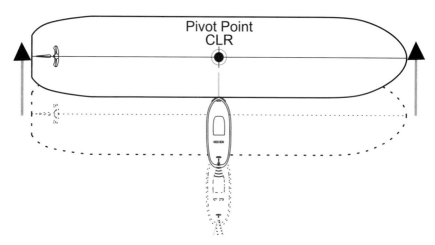

Fig. 12-5. Lateral ship movement, stopped ship with one assist tug.

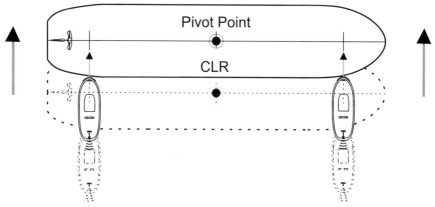

Fig. 12-6. Lateral ship movement. Stopped ship with two assist tugs

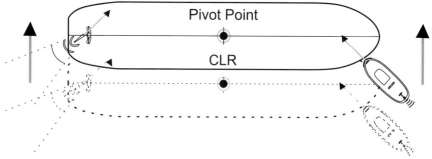

Fig. 12-7. Lateral ship movement; Stopped ship with one assist tug and ship's propulsion.

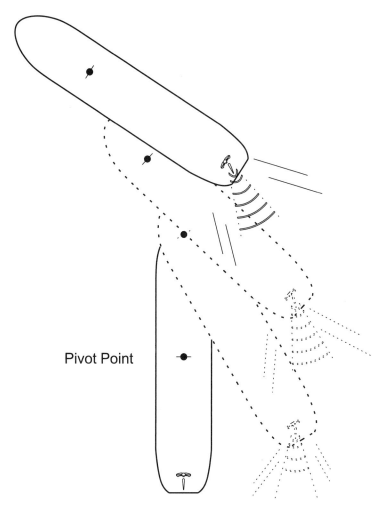

Pivot Point

Fig. 12-8. Pivot point migration on a turning ship.

Drift Angle

Drift angle is the angle between the ship's centerline and the direction in which the ship's bridge is actually traveling in a turn. When the ship has headway and is rotating around its pivot point in a turn, it is advancing forward as well. If the bridge of the ship is located well aft and away from the pivot point, the bridge is tracking sideways and forward (fig. 12-9). The shiphandler piloting from this bridge will sense the stern skidding sideways in the turn.

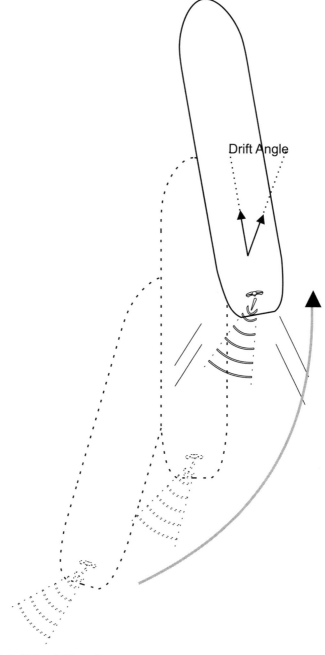

Fig. 12-9. Ships drift angle.

Ships with large drift angles typically are full-bodied ships, such as bulk carriers. Ships with finer, narrow hulls have a smaller drift angle. The difference is caused by the relationship between two factors: the momentum generated by the ship's aft section in the turn and the amount of water resistance along the hull. Drift angle serves as a visual cue for the pilot to estimate the degree of lateral motion, or slide, that a ship has as it executes a turn.

Lateral Motion

Ships turning with large drift angles have a strong lateral component. The experienced ship pilot can use this lateral element to advantage when it augments the tug's application of force.

USING TUGS TO TURN SHIPS WITH WAY

Turning or moving a ship laterally while underway requires constant vigilance on the part of the ship pilot. He must continuously assess the dynamic balance between both ship and tug as they apply force to the ship's turning lever. He must think ahead and anticipate the optimal tug position both at the start of the maneuver and throughout its duration.

A good illustration of this is the example of a ship that has backed out of a narrow waterway and is required to execute a 180° turn to head to sea. The pilot has several options, each of which is associated with a series of critical decision points (CDP).

If the tug is at the bow of the ship (fig. 12-10), the tug can initiate the turn by pushing or pulling on the bow. The tug does this to take advantage of the long assist lever to the ship's pivot point aft. One CDP is when to apply the ship's power and rudder. As the ship comes ahead with the rudder hard over, the ship loses its aft momentum and the pivot point migrates toward midships. At this point both ship and tug are working together to spin the ship within its own length. This presents another CDP. The pilot can allow either the ship to gather headway or by giving short kicks ahead, keep the way off, allowing the ship to continue turning around its midships turning axis.

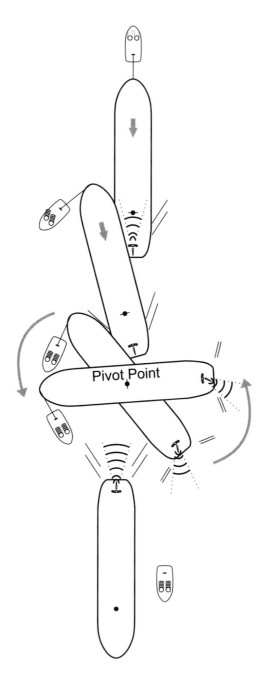

Fig. 12-10. Ship with sternway turning with one tug on bow.

If the pilot allows the ship to gather headway, the pivot point initially moves very close to the bow, enhancing the effect of the ship's rudder. As a result the ship's maneuvering lever is rotating the body of the ship around its bow and the tug is no longer effectively turning the ship. Here is yet another CDP. If the tug has enough maneuverability and power it can offer sufficient resistance to the ship's headway to hold the ship's pivot point in a position well forward. This allows the ship to continue turning with minimal headway. If the tug lacks sufficient capability, the ship continues to gather headway and the tug may be overrun by the ship unless released.

If the same situation occurs with the tug positioned at the stern of the ship, the tug cannot exert an effective turning force until the ship has taken off all its sternway. Once sternway is off and headway is gained, the tug's assist lever lengthens with the migration of the pivot point forward, becoming more efficient. The tug is now in an effective position if it is necessary for the ship to gather speed as it turns (fig. 12-11).

If two tugs are used in this turning situation, the ship pilot has many options to manipulate the location of the pivot point, the ship's advance, and its rate of turn.

USING TUGS TO LATERALLY MOVE SHIPS WITH WAY

Using tugs to control lateral movement in ships with way requires the pilot's careful balancing of sideways forces applied to the ship's turning lever by both ship and tug. The interrelation between the ship's forward inertia and the amount and location of the tug's lateral force determines the location of the ship's pivot point. One consequence of this relationship is that the tug can influence the proportion of lateral versus rotational effect on the ship. The tug does this through judicious use of its available horsepower. When a ship is moving ahead slowly with midships rudder, the tug pushing abeam of the ship's forward shoulder may cause the ship to move laterally (fig. 12-12A) or both turn and set laterally (fig. 12-12B). The degree of each effect is dependent on the amount of the tug's

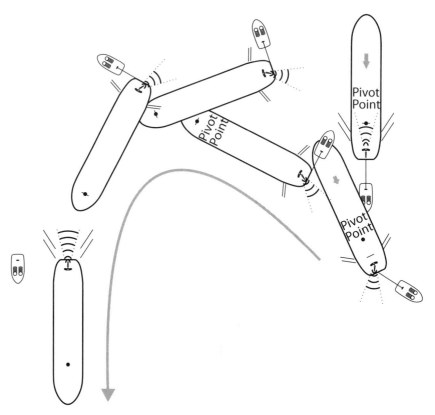

Fig. 12-11. Ship with sternway turning with one tug on stern.

applied force. This same principle applies to a ship with sternway and tug pushing on the aft quarter.

A sideways advance of the ship also can be generated by using the lateral component of a ship's turn in concert with the tug's applied force. When the ship gives a kick ahead with the rudder hard over, the ship's inertia sustains a residual lateral motion after the propeller is stopped. This lateral motion is more pronounced at the ship's stern. A tug positioned toward the ship's bow can use its force to balance or counter the swing of the stern, resulting in the desired lateral motion (fig. 12-13).

Two tugs moving a ship laterally as it advances must adjust their position and amount of thrust. These adjustments ensure the application of equal amounts of leverage on either side of the ship's

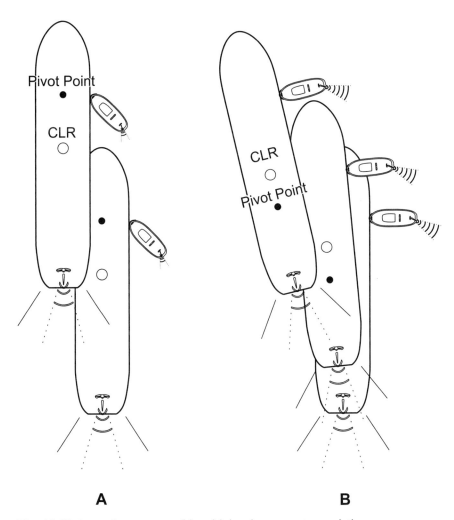

Fig. 12-12. Lateral movement-ship with headway, one tug assisting.

pivot point. This is best illustrated in the example of two tugs on towlines breasting a ship off a dock as it begins to gather headway (fig. 12-14). When the tugs first lift the ship off the dock, the ship is stopped and both tugs are pulling at equal distance from the ship's pivot point. When both tugs apply equal force, both ends of the ship come off the dock evenly and the ship keeps its original heading.

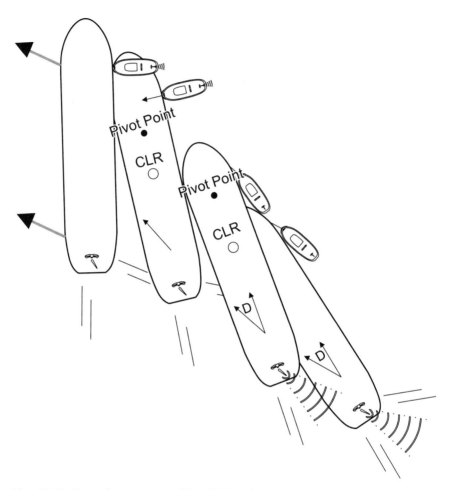

Fig. 12-13. Lateral movement, ship with headway. One tug assisting, ship kicks ahead. D: Drift angle.

Once the ship's propulsion is engaged ahead and the ship gathers headway, its pivot point shifts well forward. This immediately creates an imbalance between the two tugs. The stern tug leverage increases while the bow tug's decreases. This is a critical decision point. If both tugs continue to pull at equal force, the ship's stern begins to rotate around the pivot point and the ship begins to turn as it comes ahead. However, if the stern tug either reduces its thrust or shifts to a position closer to the ship's pivot point, it continues to balance its leverage with that of the bow tug. The ship continues to advance sideways while maintaining its original heading.

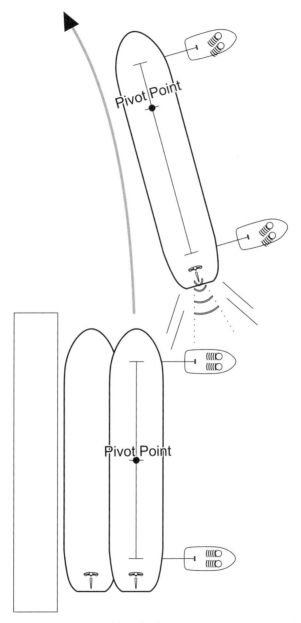

Fig. 12-14. Lateral movement; Ship with headway and two tugs assisting.

The examples illustrate the need for the shiphandler to have a comprehensive understanding of the leverage forces at work in turning and laterally moving ships. It is only within this framework of knowledge that he can position tugs to control the substantially larger mass of the ship. This may be compared to a small judo master using imbalance and leverage to throw a much larger opponent to the ground. Indeed, the shiphandler who skillfully uses a small tug to turn and shift his ship, exhibits a similar mastery of his art.

STUDY QUESTIONS

1. Is the pivot point always near the ship's center of lateral resistance?
2. What is the effect on the ship's pivot point of one tug pushing at the bow?
3. How many ship lengths will it take to turn a ship with one tug pushing at the bow?
4. What is the effect on the ship's pivot point of two tugs pushing with equal force, one on the port bow and one on the port stern of a stopped ship?
5. In order to move a stopped ship laterally the external forces applied must have_____ on either side of the ship's pivot point.
6. What is drift angle?
7. Why is knowledge of drift angle useful to the pilot?

CHAPTER THIRTEEN

CHECKING A SHIP'S WAY WITH TUGS

A prime function of tugs in shipwork is to create a braking effect on the ship's forward momentum. Speed management of ships plays a critical role in the safety, efficiency, and economics of bringing ships in and out of berths. Previous chapters in this book have described the numerous negative effects excessive ship speed may have on tugs.

In addition to the safety considerations for the tug, ship speed also may have a negative effect on the ship's own steering and propulsion. The ship's rudder requires a minimum amount of water flow to be effective. The velocity of this flow is the combined result of ship speed and propeller wash. As ship speed and propeller revolutions drop, so does the pilot's control of the ship. If ship speed drops too low the pilot can lose control of the ship unless assist tugs are present.

The inertia of large displacement ships is quite powerful, even at slow speeds. Many ships have poor backing qualities and can require a considerable distance to come to a stop under their own power. A 60,000 DWT loaded Panamax ship traveling at three knots may need over a half-mile to come to a stop. While ship's engines are backing over this distance, the pilot has little or no control of the ship in the absence of tugs.

When tugs are employed to manage ship speed, the pilot can use the tug's braking force in two ways. One is simply to assist in bringing the ship to a stop. The other is to provide sufficient drag to keep ship speed down without actually bringing it to a stop. This practice is particularly useful on ships where dead slow is relatively high (five to seven knots). The tug's braking effect allows the pilot to continue to drive the ship with the ship's rudder and propeller. He can control the ship and keep its speed down to a reasonable level while maneuvering in waterways and approaching berths.

From a shiphandling perspective, ship speed represents a compromise. This compromise maintains the balance between tug safety, control of the ship, and a healthy respect for the power of the ship's inertia. However, there are commercial considerations to managing ship speed.

Time is money in marine transportation. Ship charter rates are tens of thousands of dollars per day. Dockage fees and stevedoring rates add to the cost of ships in port. Pilots are under considerable pressure to bring ships into berths on time, every time. Ship owners and agents have no inhibition in voicing their frustration when watching their ship take hours to creep into a berth while the longshore crew stands on the dock, being paid to wait.

The ship pilot must account for all of these factors in his management of ship speed. The tug is one of the primary tools to assist him in this task.

TUG POSITION

The tug can apply a braking force to the ship either while alongside the ship or in-line aft. When the tug is alongside the ship it lays offset from the ship's centerline. Any braking force will tend to cant the ship toward the tug side of the ship (fig. 13-1A). This effect can be negated by having the tug in-line aft (fig. 13-1B) or by employing two tugs alongside, each on opposite sides of the ship (fig. 13-1C).

BRAKING TECHNIQUES

The tug can apply a braking force to the ship by using the tug's weight, lateral resistance, bollard pull, or a combination of these techniques.

Tug's Weight and Lateral Resistance

The tug laying passively alongside being pulled along on a headline or towline, creates a drag force. Tugs positioned aft of the ship on a

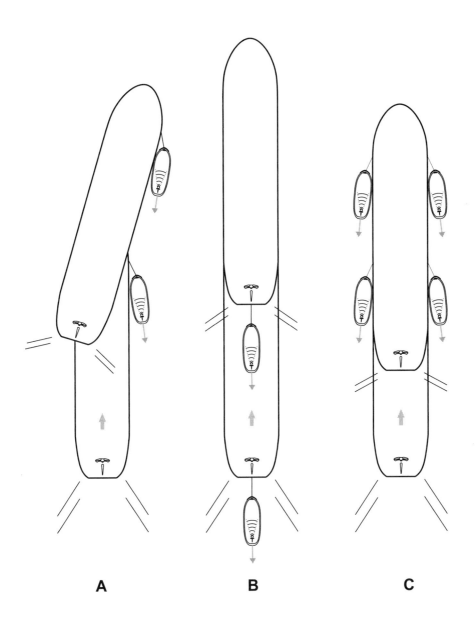

A   B   C

Fig. 13-1. Effect of tug(s) backing on ship. A: One tug alongside. B: One tug in-line aft. C: Two tugs alongside.

headline or towline can increase this drag by turning the tug's hull more broadside to the ship's direction of advance. A single-screw conventional tug on a towline typically uses a gob line. A twin-screw tug may be able to twist into this position with engine power alone (fig. 13-2A). The tractor or ASD tug can produce the same effect by going into an indirect towing mode (fig. 13-2B). These are good techniques to use when the pilot desires the consistent effect of a drogue to assist in the management of ship speed.

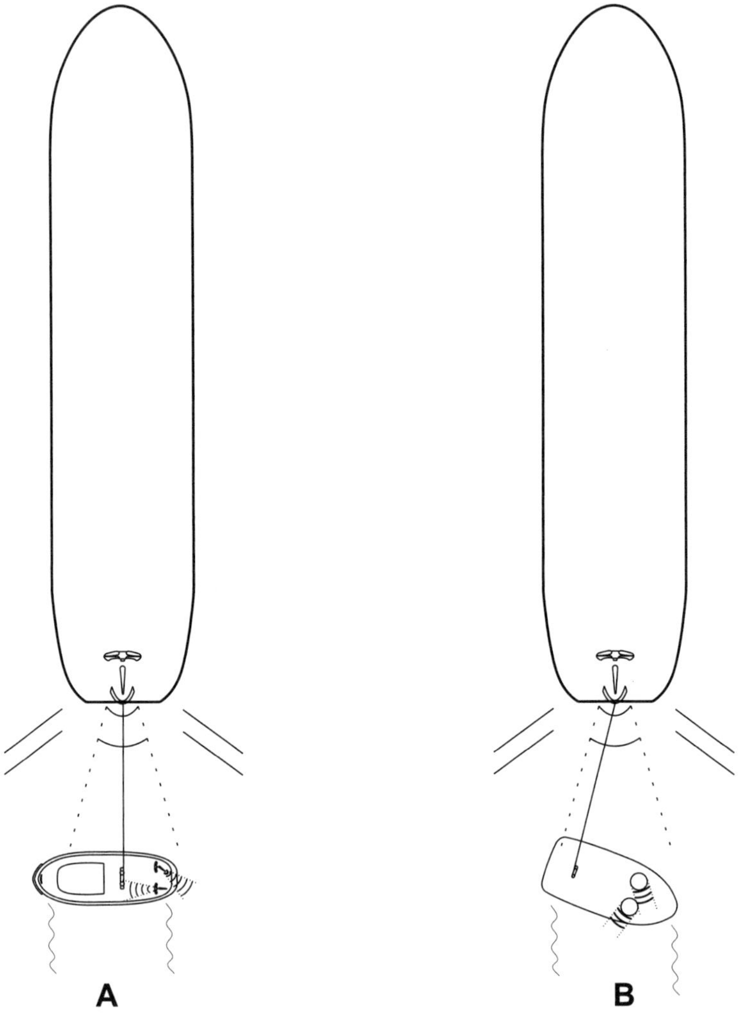

Fig. 13-2. Tug's drag as braking force. A: Conventional tug. B: Tractor tug.

Tug's Bollard Pull

Of course the tug's bollard pull also can be applied to retard a ship's advance. A conventional tug can back on a headline or, at slower speeds, tow on a towline. The tractor or ASD tug can employ an in-line, reverse arrest to create sufficient braking forces (fig. 13-3A).

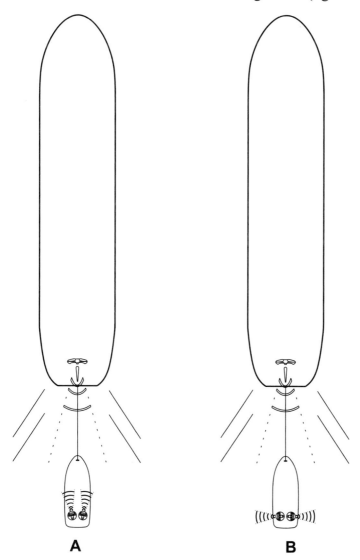

Fig. 13-3. Tug's bollard pull as braking force. A: Reverse arrest. B: Transverse arrest.

Transverse Arrest

The ASD tug has an additional option of using the transverse arrest position (fig. 13-3B). This technique is commonly employed at higher ship speeds and is particularly effective in ship escort work.

CONSIDERATIONS

The ship pilot's choice of tug position and technique must consider several factors that affect a tug's braking capability.

- ship speed
- negative water flow
- ship's propeller wash

Ship Speed

Ship speed has a direct bearing on line load to the tug. Even before the tug applies bollard pull to its line, the line is preloaded with tension resulting from the tug's hydrodynamic drag. The amount of this drag is a function of the square of ship speed. Excessive speed may use up a critical amount of the line's breaking strength. Too much way on the ship may restrict the tug from applying full power if needed, or cause the line to part under additional load.

Negative Water Flow

When a tug reverses its thrust to slow a ship, the tug's propulsion machinery must contend with the effects of negative water flow. High propeller loads may overload and stall the engines of conventional tugs. The same effect can occur on tractor tugs using too much pitch.

The ASD or tractor tug with independently controlled thrusters can manage this effect by turning its units at an angle to the tug's centerline. The increase of the azimuth angle reduces the effects of negative water flow. The tug operator can choose the optimal angle due to the independent control of the azimuthing units. He can align the units anywhere between a reverse-arrest (units at 180° to water flow) and a transverse arrest.

Ship's Propeller Wash

The braking tug positioned aft and in-line with the ship's centerline operates in the ship's wake and propeller slip stream. This places the tug in turbulent water that may cause the tug's propulsion to cavitate and lose efficiency. In addition, the varied velocities of water flow, combined with eddies associated with this area behind the ship's stern, make it difficult to hold the tug in position.

The conventional and VSP tractor tug may find it difficult to apply a consistent, direct pull on the towline while keeping the tug in this position. The transverse arrest technique works well in these situations, both holding the tug in position and applying a braking force. This technique is available only to the ASD, or the tractor tug with independent steerable propellers.

Getting a ship to stop at a predetermined point can be a daunting task. It is a task that is rife with opportunities for failure. The helmsman may not understand the pilot's order for "back half." The ship's engineer may not respond in a timely manner. The ship's propulsion may malfunction. The propeller may cavitate instead of bite into good water. An unpredictable following current may propel the ship forward. All the while the ship's inertia continues to carry it toward the dock, bridge abutment, or other immovable objects. The pilot's expression may not betray his inner doubts, but the same question always arises: "Will I get this ship to stop in time?"

Many times tugs are called on to help remedy mechanical mishaps and human misjudgment. Tugs are an effective and useful tool if employed properly. But to ensure they are effective at these critical times, the pilot must have prepositioned the tug in a manner that enables the tug to employ its most effective braking factors associated with the design and propulsion features of the tug.

STUDY QUESTIONS

1. What are the three principal techniques used by tugs to apply a braking force to the ship?

CHAPTER FOURTEEN

BASIC SHIPHANDLING MANEUVERS WITH TUGS

There may be as many ways to handle a ship with tugs as there are ship pilots. A pilot's shiphandling style is merely a point in time on the pilot's continuum of learning. His individual style is based on the techniques that have proven successful for him, he has seen work for others, and learned from the experience and knowledge shared by the professional mariners who served as his mentors. As one pilot states:

> "We each come into the pilots after years as a captain and bring our varying experiences. Only after getting the pilot's license and working as a pilot do we develop into a pilot. Most of us take years to learn from our individual experiences and talking to other pilots to evolve in what we eventually become."

The ship maneuvering techniques illustrated in this chapter are offered from that perspective. They are the products of the collective experience among professional mariners and are presented to illustrate the principles of successful shiphandling techniques with tugs. However, they are by no means completely inclusive and certainly should not be taken as the one and only right way to maneuver ships with tugs. No two dockings are ever exactly the same. The variations in weather conditions, berths, ships, tugs, and the humans that command them continually conspire to challenge ship pilots day by day and maneuver by maneuver.

Learning how to use tug assist in basic ship maneuvering situations is a staple of a pilot's training.

A principal axiom of maneuvering ships with tugs is to "drive the ship". Trainees are consistently reminded by their more experienced mentors to "drive the ship, assist with tugs." Unless it is a dead ship, the tugs are there to augment the ship's propulsion and maneuvering system, not supersede it. This is an essential perspective for viewing the illustrations that follow.

During any maneuvering sequence ship pilots are engaged in multiple tasks. They are constantly processing, filtering, and prioritizing information while making judgments, active decisions, and communicating directions to ship personnel and tug. All this while continuing to drive the ship and maintain a feel for its behavior.

Although ship pilots have individual styles, their approaches to a ship maneuver and its associated tasks share common elements. In broad terms, this process has three phases. The first is an initial assessment of:

- The ship
  - o   displacement condition
  - o   handling characteristics
  - o   identifying the functionality and reliability of the ship's bridge resources
- Human resources (master, mates, helmsman)
- Electronic resources (communication and navigation equipment)
- Communication reliability (bridge to engine room)
- The weather
  - o   wind and current
  - o   visibility
- Navigation conditions
  - o   water depths
  - o   channel constrictions
  - o   air draft restrictions
  - o   berth configuration
  - o   vessel traffic conditions

- The tugs
  - o hp, maneuverability and bollard pull
  - o limits and capabilities
  - o experience and skill of tug operators

In the second phase, the pilot uses the information gathered in the initial assessment and creates a maneuvering plan. The plan includes:

- the ship's intended track to or from berth
- the ship's intended speed
- the placement of tugs
- identifying critical decision points (e.g., helm over points, points to initiate speed reduction procedures etc.)

The third phase is plan implementation and adjustment. Just as no two dockings are alike, no maneuver unfolds exactly as planned. The ship pilot must constantly reassess and revise the maneuvering plan. He is continually:

- sensing ship movement and changes in environmental influences
- planning future maneuvering steps, aware that his choice of a maneuver at any one moment may restrict his choices later
- managing communication and trust relationships with the ship's master, crew and tug operator
- determining the best use of the assist tug's application

His directives to the ship's bridge and the tug must keep pace with the many subtle, sometimes dramatic, changes in the initial maneuvering conditions. A pilot's ability to quickly sense these changes—the gust of wind, the drift of the ship's bow—and make quick and accurate adjustments is the mark of a skilled and experienced pilot.

Placement of Tugs in Shipwork

There is neither an absolute formula nor a standard governing the optimal placement of tugs in shipwork. There are guidance principles, however, which can assist the pilot in his decisions.

1. Anticipate the complete sequence of maneuvering events in which tugs will be used:
   a. Note the points in the maneuver where the tug may have to switch position, transition from towing to a push/pull mode.
   b. Note where the tug's push or pull will be critical to the success of the maneuver.
   c. Place tugs in positions based on the priority and criticality of their use.

2. Factor in the safety of the tug:
   a. Avoid positions that place the tug between the ship and the dock or other fixed objects with no means of escape.

3. Factor in the limits and capabilities of specific tug designs:
   a. Put the most maneuverable tug in the position that requires the tug to shift.
   b. Place a tractor or ASD tug in positions where a conventional tug would primarily have to back in order to be effective.
   c. Use handy tugs (those with a low profile) to work near the ends of ships and under overhanging structures.

4. Factor in the skill level of the tug operator.

5. When multiple tugs are used, position the most powerful tug where its horsepower is most needed:
   a. This may be forward on a deep-loaded ship, or
   b. aft on a ship in ballast.

6. Estimate the amount of the tug's horsepower needed to keep the tug in the assigned positions versus the amount available to push or pull on the ship.

7. On large, deep-draft vessels, tugs are most effective working as far forward or aft as is safe or practicable.

8. On smaller vessels or those of light draft forward, the bow tug may function better if secured farther aft than would be proper for a larger or deeper vessel.

Clearly, positioning a tug in shipwork is a judgment call. If the shiphandler has any doubts about the appropriate placement of the tug, he should solicit the expertise of the tug captain on the safest and most effective tug position.

The tug's position may seem inconsequential to the casual observer. However, people familiar with shipwork understand that securing the tug a few feet farther forward or aft or at the angle at which a tug lies when it pushes or pulls can have a significant effect on how the ship responds. Poor placement of assist tugs may be more than just ineffective, it may be a detriment.

The diagrams that follow illustrate some basic ship maneuvers with tugs. Certainly, diagrams in a book cannot convey the feel or sense of handling a ship. However, they can provide insight into the rationale for placing tugs in particular configurations and are an excellent tool to highlight the application of the principles involved with shiphandling with tugs.

CONDITIONS: NO WIND OR CURRENT

Docking Maneuvers

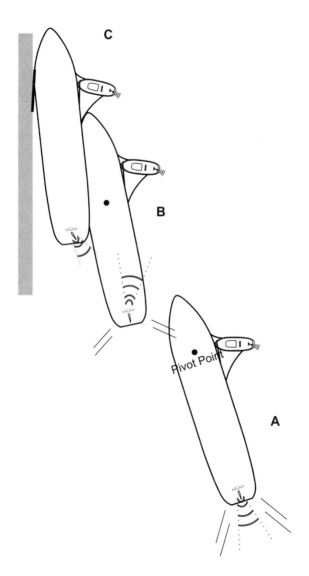

Fig. 14-1. Docking, one tug, push/pull. A: tug is pushing on pivot point. B: Ship backs, propeller torque moves ship's stern towards dock, ship's pivot point moves towards amidships. C: Tug breasts ships bow to dock, springline run and ship works on springline to swing in stern.

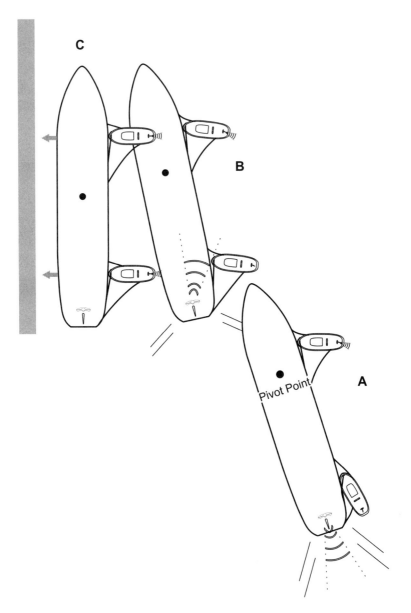

Fig. 14-2. Docking, two tugs, push pull. A: Forward tug pushing on pivot point, after tug stopped. B: Ship backs, forward tug pushing on pivot point, after tug stopped. C: Ship stopped, pivot point amidships, both tugs pushing equally. Note this is a good example of "Driving the Ship", even with the presence of the after tug the approach is the same as in fig. 14-1.

The aft tug is used to breast the ship's stern in place of the springline.

Undocking Maneuvers

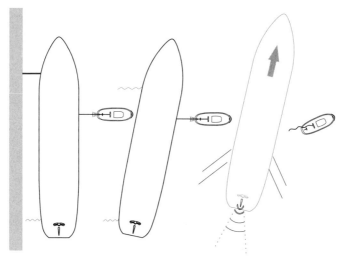

Fig. 14-3. Lift off pier, one tug pulling forward of midships. Breast line from ship is held and slacked as needed to lift the ship evenly off the dock. Once the ship's stern is clear, tug is cast off.

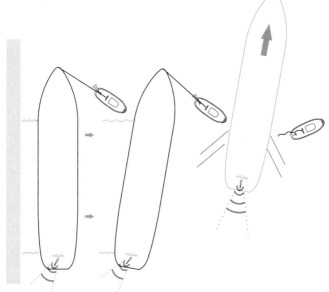

Fig. 14-4. Lift off pier, one tug pulling on a towline fast to the bow, working against the ship's engines and rudder.

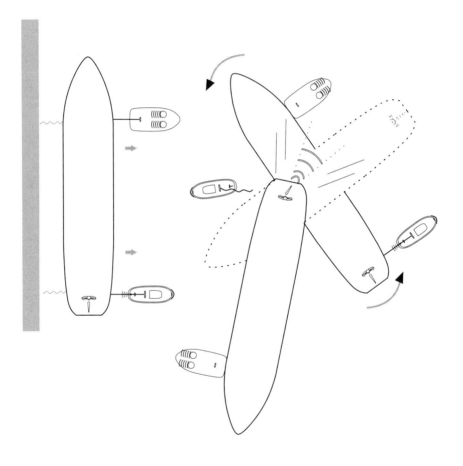

Fig. 14-5. Lift off pier, turn ship 180 degrees, two tugs. Tractor/ASD tug is positioned on bow, as it can easily transition to push on ship's starboard bow. Conventional tug can stay on towline until ship comes ahead.

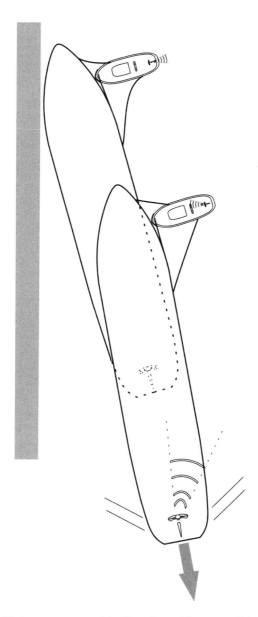

Fig. 14-6. Lift off pier, one tug push/pull on bow. The tug will breast the bow of the ship to the dock and open up the stern. The tug can then steer the ship as it backs clear of the dock. A conventional tug may need a quarter line.

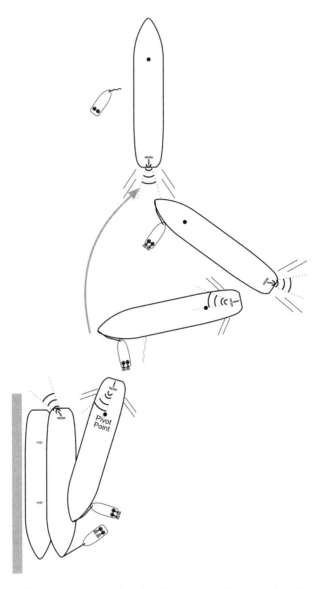

Fig. 14-7. Docking a vessel at a berth with current ahead setting the vessel on to the dock. The tractor/ASD tug keeps the bow from setting too heavily on the pier. A second tug may be used on the offshore quarter aft, in the event the ship's steering and propulsion are not sufficient to control the stern while docking.

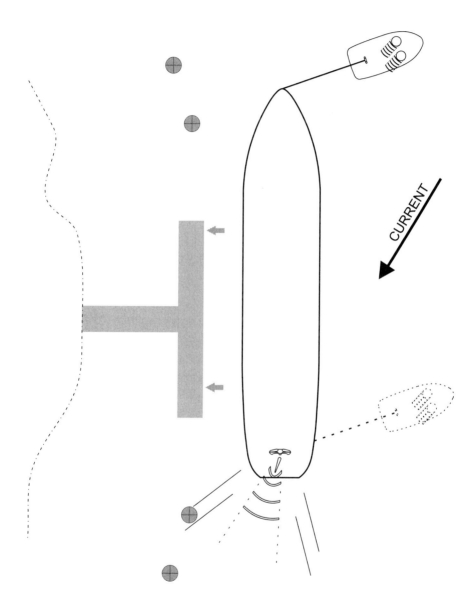

Fig. 14-8. Lift off pier, turn 180 degrees, one tug. Tractor/ASD backs on bow working against ship's engines and rudder. As ship clears dock tug pushes on bow while pivot point is well aft. Ship comes ahead with right rudder to check sternway and turn ship.

CONDITIONS: WIND OR CURRENT SETS ONTO DOCK

Docking Maneuvers

It is sometimes convenient for the tug to work on the inboard side of a ship when undocking. This is done most commonly at tanker berths where openings between piling clusters leave enough space for the tug to work safely (fig. 14-9).

When a vessel is undocking, a tug may be stationed on the inshore side. The tug stations itself where it can work safely between the dolphins, and then breasts the ship clear of the dock.

Sometimes a tug must shift from forward to aft or from one side of the ship to the other. To avoid delays, it is preferable to assign the most maneuverable tug to that position, unless other considerations such as horsepower make it impractical.

The shiphandler often is involved in a fairly intricate maneuver, but his concentration should not cause him to overlook factors that can affect the tug and ship. If a ship is being turned through the wind and is light, the shiphandler may want to shift the tug to the other bow of the ship. Once the bow has passed through the eye of the wind, the tug should back enough to check the vessel's swing before letting go. This gives the tug time to shift to the new lee bow without having to race around to the other side. Another hazard is the possibility of the tug landing heavily on the new lee bow due to the speed of the ship's turn.

If a tug must shift from one quarter to the other, it is helpful if the ship can avoid backing hard until the tug is fast. Otherwise the ship's wheel wash might carry the tug too far forward. This also can delay getting into position to work and lead to an accident. Some apprentice shiphandlers have difficulty handling transitions from steering to turning, moving the vessel laterally, etc. In this case, it is best to break the maneuvers down into steps, such as:

- stop the ship
- turn the ship
- shift tugs
- breast vessel to dock

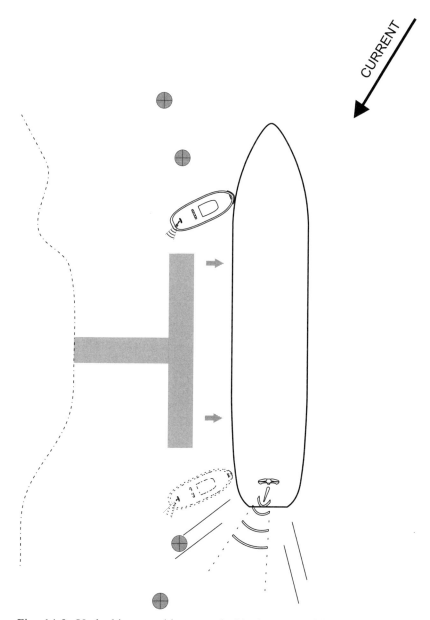

Fig. 14-9. Undocking, working tugs inside between piling clusters. A vessel undocking at a typical tanker berth with current from ahead, setting the vessel on the dock. A tug can be employed effectively on the inboard bow to breast the bow away from the dock. A tug may be used on the inshore quarter aft, in the event the ship cannot lift its stern off the dock using its own steering and propulsion.

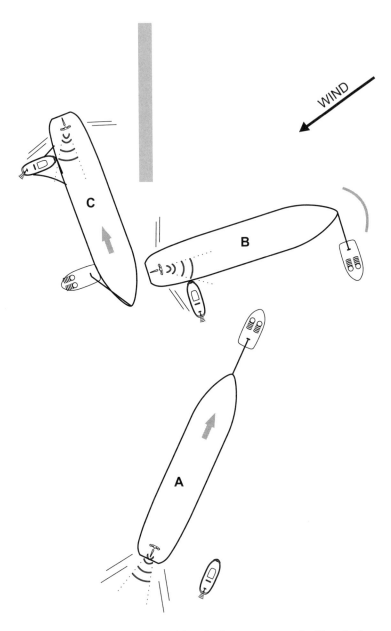

Fig. 14-10. Docking, stern to/lee side of pier, two tugs. A: Vessel slows and conventional tug aft shifts to starboard quarter. B: Vessel stops and tractor tug shifts to push/pull starboard bow (pulls bow). C: Vessel backs easy until it is in position with tugs assisting as needed.

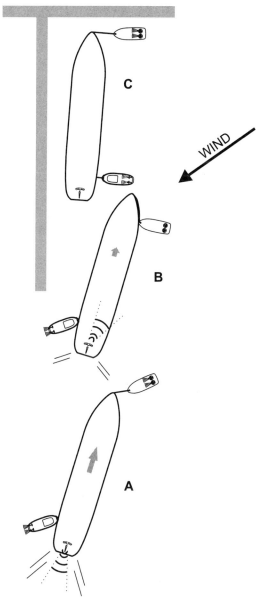

Fig. 14-11. Docking, bow in, windward side of finger pier, two tugs, tractor/ASD up forward, conventional astern. Position A: Ship is proceeding slowly ahead with forward tug holding bow up towards wind. Position B: Forward tug has swung around starboard bow to push or pull as required. The after tug lets go its line, but continues to push as needed on the port quarter as long as it can stay alongside the ship. Position C: The ship is now dead in the water, with both tugs fast on the starboard side. Tugs can push/pull to berth the vessel.

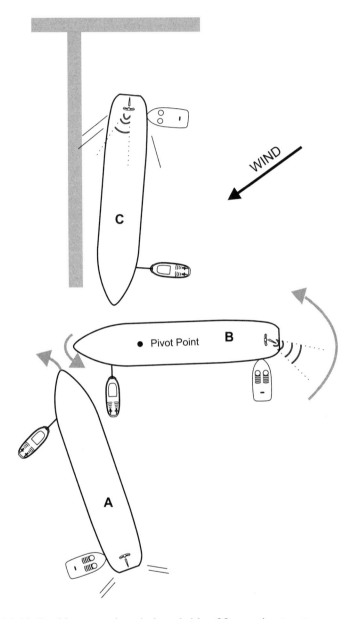

Fig. 14-12. Docking, stern in, windward side of finger pier, two tugs, conventional tug on bow, Tractor/ASD on stern. A: Ship stops, tugs push/pull to turn ship. B: Aft tug pushes harder than forward tug pulls to keep pivot point forward. C: Ship maneuvers with its engine ahead and astern, tugs push/pull as required. Note: Conventional tug may require quarter line.

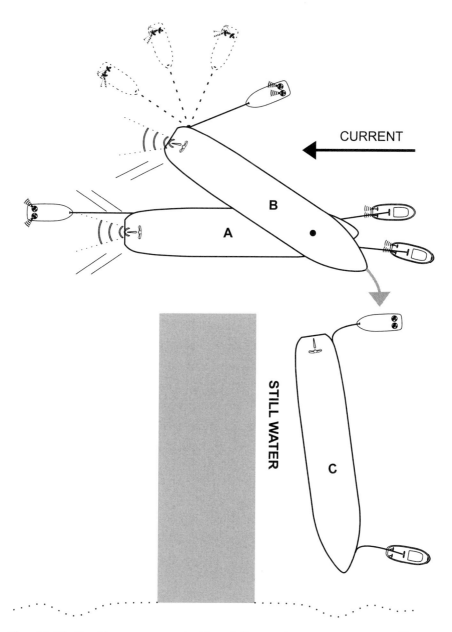

Fig. 14-13. Docking, up current side of finger pier, two tugs on towlines, conventional tug forward, tractor/ASD aft. A: Ship stemming current, small amount of headway. B: Initiate turn, aft tug must shift to pull up current. C: Tugs pull as needed to allow ship to settle onto dock. Note: Tractor/ASD tug is positioned aft to use maneuverability to shift up current in position B.

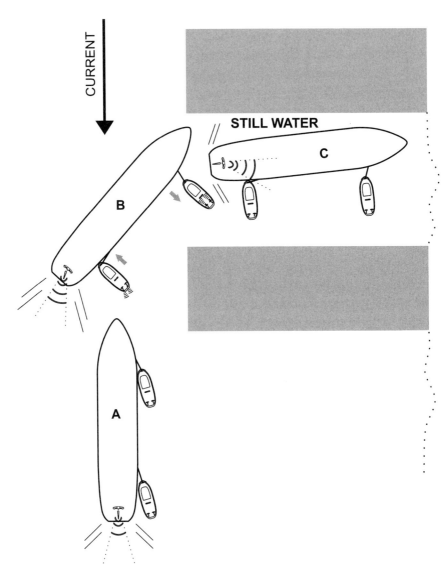

Fig. 14-14. Docking, bow in, down-current side of pier, two tugs. A: Ship stemming current, small amount of headway, both tugs alongside, ready to work. B: Initiate turn, forward tug may need to back when bow enters still water. C: Tug's push/pull to breast ship into pier.

Undocking

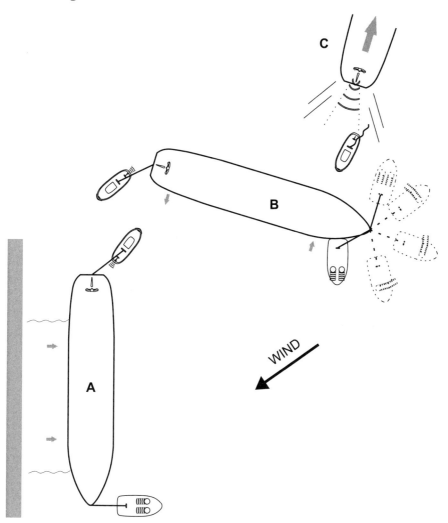

Fig. 14-15. Undocking, lift ship off weather berth, two tugs on towlines, tractor/ASD tug up forward. Position A: Tugs lift ship off pier laterally and astern. Position B: Forward tug shifts to push on ship's starboard bow. Position C: Both tugs released, ship underway. Note that the Tractor/ASD is positioned up forward, as it can transition to push much quicker than conventional tug. When letting tugs go, pilot should allow enough time for aft tug's tow hawser to be brought onboard prior to engaging ship's propeller.

CONDITIONS: WIND OR CURRENT SETS PARALLEL TO DOCK

Docking

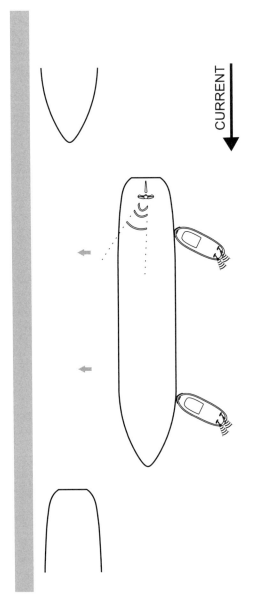

Fig. 14-16. Breasting into berth with fair current, two tugs. Note ship backs as needed to stem current.

Fig. 14-17. Docking, head current, one tug on the hip. Note that tug coordinates with ship's propulsion for twin-screw effect.

Undocking

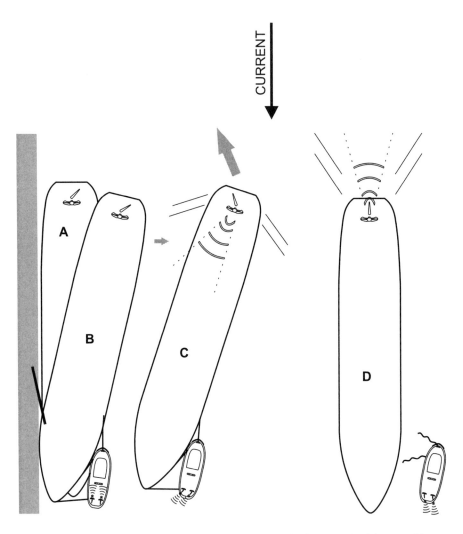

Fig. 14-18. Undocking, stern to current, one tug at the bow on the hip. A: Ship's rudder is turned to port. Stern has sternway relative to the water flow lifting stern off dock. B: Current gets between ship and the berth, striking ship along its starboard side, moving ship bodily away from berth. C: Tug comes ahead with left rudder to move bow away from berth, ship shifts rudder, run engines astern to stem the current. D: Ship is straight in channel and tug is released.

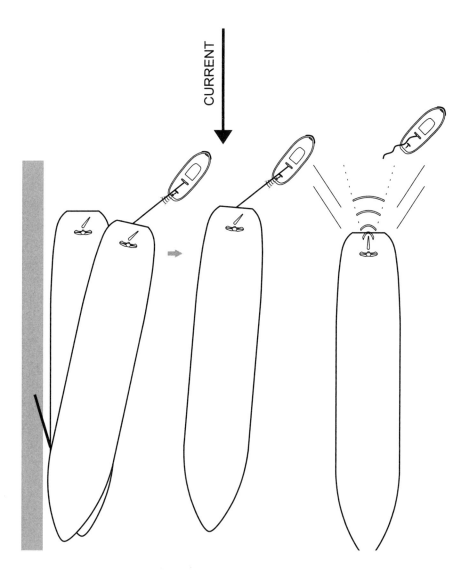

Fig. 14-19. Undocking, stern to current, one tug on towline.

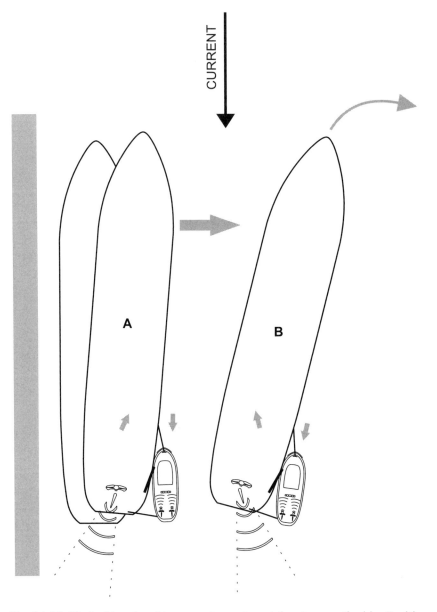

Fig. 14-20. Undocking, head to current, one tug at the stern on the hip. Position A: Ship twin-screws with tug to open bow. Position B: Ship shifts rudder to turn ship to starboard.

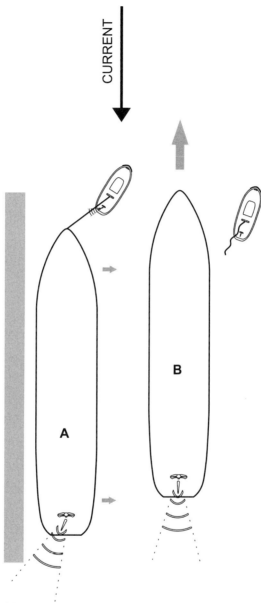

Fig. 14-21. Undocking, head to current, one tug on bow on a towline. Position A: Tug pulls bow out while ship comes ahead with left rudder to stem current and move ship laterally away from berth. Position B: Ship straight in stream, tug let go. Note the tug on the towline must keep its bow up into the current to avoid being swept down alongside the ship.

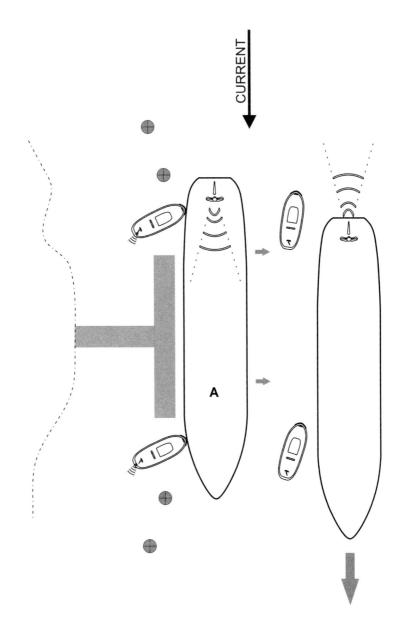

Fig. 14-22. Undocking, stern to current, two tugs working inside, no lines. A: Ship moves laterally off dock as tugs push while ship backs to stem current. Note that tugs do not put up lines so they can clear out quickly once the ship is in stream.

MAKING A BEND

Ships approaching a bend may have tugs on either side of the bow to facilitate controlling the ship. If the vessel is light forward, the tugs can be useful even if the ship is moving fairly fast (five to six knots). If deep laden, the ship should move slow enough (three to four knots) to enable the tug on the outboard side to push at an angle of 45 degrees or more to the ship (fig. 14-23A). Otherwise the tug will have little turning effect and may increase the ship's speed.

A tug on the inside of the bend can be used effectively backing (fig. 14-23B). This cants the bow in the direction of turn, and resistance of the tug retards the vessel's advance and makes the ship's rudder effective. Ship's speed is critical to the tug in this case. If the ship is traveling too fast, the weight of the tug being towed alongside may impose such a severe strain on the headline that it might part if the tug backs hard.

On the outboard quarter of a ship, a tug can do very little unless it is hipped to the ship. In this case the tug can come ahead with its helm turned toward the vessel (fig. 14-23C). If the ship backs, it has a twin-screw effect.

A tug on the inboard quarter can assist by backing if it is hipped up (fig. 14-23D), or by pushing if it has only a headline up (fig. 14-23E). However, it will be more effective pushing if the bend is sharp. Here again the vessel's speed should be slow enough for a tug to maintain at least a 45 degree angle to the ship.

NARROW BRIDGE OR LOCK TRANSITS

One of the most demanding shiphandling maneuvers with tugs is to transit through a narrow bridge or lock. The tugs must be able to respond quickly and precisely to the ship pilot's directions. Tugs also must be able to apply steering, propelling, or braking forces to the ship within a distance that may not be much wider than the ship's beam. Most of these types of transits call for a lead tug on a towline and a trailing tug.

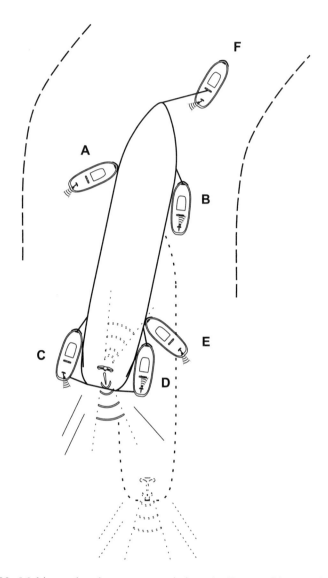

Fig 14-23. Making a bend, one tug assisting. A: Tug pushing at the bow on outboard side of ship's turn. B: Tug backing on line at the bow on inboard side of ship's turn. C: Tug hipped up on the outboard side of ship's turn. The ship twin-screws around bend with tug coming ahead with rudder over and ship backing. D: Tug hipped up on the inboard side of ship's turn. The ship twin-screws around by tug backing and ship coming ahead with rudder over. Position E: Tug pushing at the quarter on inboard side of ship's turn. F: Tug on towline. Note that on very hard bends, when the tug must pull at a sharp angle, it is sometimes necessary to back the ship's engine to avoid tripping the tug by overtaking it.

The lead tug assists in steering the bow (or stern if going stern first) and provides a forward-propelling force. The trailing tug's function is to steer the stern and provide a retarding force to the ship. In practice, the vessel is maneuvered through the bridge much like a hook and ladder fire truck when it transits narrow city streets. One fireman steers the rear of the truck, while the other steers the front from the cab.

For such close quarters work, the lead tug must be made up with a very short towline and bridles may be needed. Gate lines also might prove useful for this type of work. Gate lines are separate lines led from either side of the bow to the opposite quarter of the tug.

Illustrated below are some common configurations using different combinations of tugs.

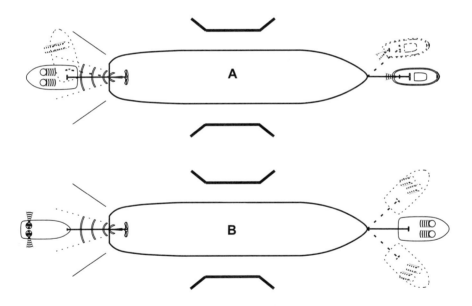

Fig.14-24. Bridge transits. A: Two tugs. A conventional tug on bow, a tractor tug on stern. Conventional tug was chosen as the bow tug for its ability to have an effective high towline angle and not extend much outside the side shell of the ship. The tractor tug was assigned aft due to its multiple options for braking (powered pull on line, lying indirect as ship drags tug) and its option to steer through direct and in-direct towing. B: Two tugs, tractor on bow, ASD on stern. Note the tractor was assigned to the bow due to its lateral mobility. The ASD was assigned to the stern due its option of using a transverse arrest to retard ship speed.

MANEUVERING LARGE VESSELS

Moving a large vessel may take several tugs to provide the total bollard pull and the appropriate tug configuration. There must be enough total bollard pull to counter the effects of wind, current, and the ship's inertia. The total bollard pull must be distributed among enough tugs to give the pilot flexibility to locate the tug's applied force at multiple points along the ship's hull (fig. 14-25). One tug may be appropriate for a 20,000 DWT ship, but four or more may be necessary for a ship of 300,000 DWT. Although experience is the pilot's best and most common resource in determining the numbers and placement of tugs, he also should use simulations and engineered calculations when available.

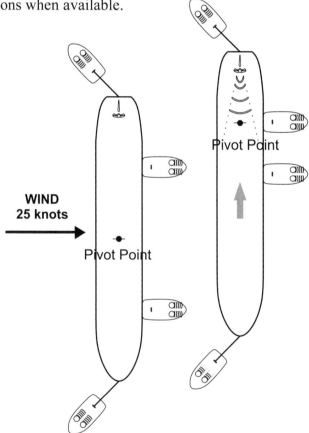

Fig. 14-25. Holding large ship broadside to twenty-five knot cross wind, four tugs. A: Ship is stopped, pivot point is amidships, tug force distributed equally along ship's hull. B: Ship making sternway, pivot point aft, tug force redistributed.

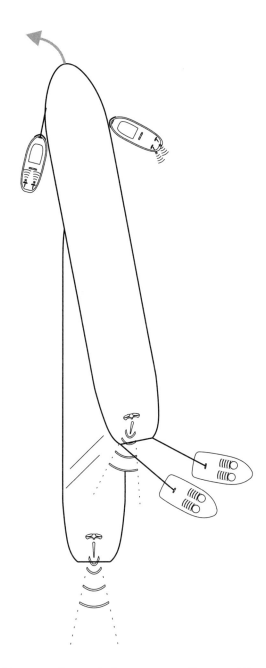

Fig. 14-26. Large ship proceeding under its own power, four tugs. The conventional tugs forward assist in steering. Tugs aft are tractor or ASD tugs due to their superior ability to steer and brake the ship. Stern tugs should have equal length towlines.

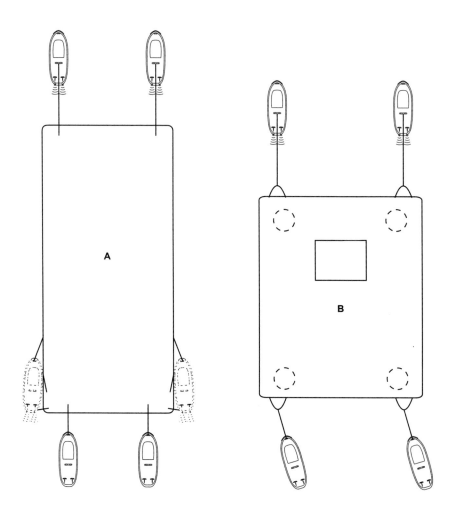

Fig. 14-27. Assisting large, unpowered vessels, four tugs. The vessel is a large barge of the type used to transport offshore drill rigs. The after tugs may hip up to the barge in protected waters for better control if the structural limitations permit.

STUDY QUESTIONS

1. What are the three phases of ship handling task management in maneuvering ships with tugs?
2. What are the factors in choosing the placement of tugs in shipwork?

CHAPTER FIFTEEN

HANDLING DEAD SHIPS

The dynamics of towing a dead ship are quite different from handling a powered vessel, and a wide range of tug configurations are possible. In dead ship towing the tugs provide both the steering and propelling forces to maneuver the ship. The tug applies steering force through a combination of power and leverage. In steering, a smaller tug can take advantage of leverage to compensate for its size. There is no such aid in propelling a ship. It is a simple matter of enough bollard pull to accelerate, retard and, most importantly, stop the ship's mass. On large ships the total amount of required bollard pull can be reached only by employing multiple tugs.

There are three common ways of handling dead ships:

- tugs alongside
- tugs on towlines
- tug alongside and lead boat

TUGS ALONGSIDE

One Tug

A small ship may require only one tug, made up for breasted towing. Tugs handle quite sizeable barges unassisted and can handle a ship of comparable size just as well. If the pilot has any doubts about where the tug ought to make fast, he should consider three factors:
- the required ship orientation at the destination dock
- the types of turns in route
- the trim of the ship

The assessment of these factors help the pilot determine which end of the ship is optimal for the tug to be made up to.

One aspect of the first consideration, ship orientation at the destination dock, is obvious—the tug will be caught between the ship and pier if it is made up on the ship's side that will be next to the dock. Another aspect can dismay the inexperienced shiphandler. Due to the offset position of the tug, the ship may dive toward the dock when the tug backs. If the pilot has misjudged the distance required to stop, the ship strikes the dock before it has lost sternway (fig. 15-1).

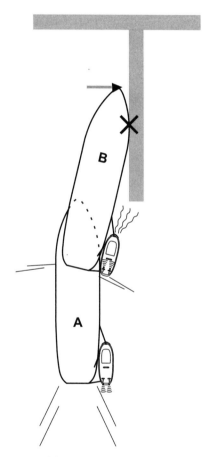

Fig. 15-1. Dead ship approaching dock, one tug, made up on pier side. Position A: Tug is made up on pier side of ship. The ship's speed is too high. Position B: Tug backs, ship sheers toward and strikes dock before headway is lost.

The ship's spotting orders at the destination dock also dictate whether the ship will lay head in or stern in, and how close it will be to other objects secured to the pier. It is easier for a single tug with a ship on the hip to approach a dock with headway, as opposed to backing in. The exception to this is if the tug is a tractor or ASD tug. Their maneuverability astern may give them the added option of backing into a berth if space allows. As an example, consider a dead ship move that ends at a finger pier, stern in, port side to (fig. 15-2). Making up to the ship's starboard bow is the best option.

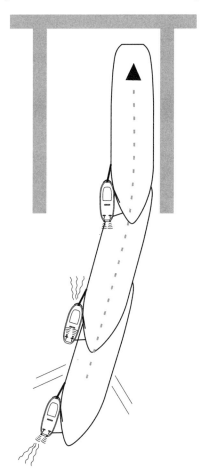

Fig. 15-2. Dead ship to finger pier, stern in, port side to, one tug.

The types of turns the ship must make also should influence the pilot's choice of tug placement. If a hard turn is anticipated it will be best if the tug made up on the hip is located on the inboard side of a turn (especially if it is a single screw). The reason for this is if the tug backs its engine it will accelerate the swing without adding headway to the tow, and can back and fill as necessary even in a very sharp turn or confined quarters without losing control of the tow. However, if a tug is on the outboard side of a turn and backs, it counters the swing and must continue to power throughout the turn. When turning in close proximity to a bank or dock, misjudgment in these circumstances can lead to a point of no return. This occurs when the ship does not have the turn rate to make the turn, lacks the space to allow the ship's trailing end to rotate around its pivot point, or has insufficient distance ahead of the ship to stop before impacting the bank or dock. If the tug backs down, it only reduces the impact of an untenable situation (fig. 15-3).

Taking the ship's trim into consideration, it may be just as convenient to move the vessel stern first as it is to move it bow first, though a little more care may be necessary in docking stern first to avoid damage to the rudder and the propeller. The heavier end, whether bow or stern, generally has more directional stability than the lighter end. Placing the tug on the lighter end gives the tug better control over the tow when it is moved heavy end first. This places the ship's pivot point near the heavy end and helps reduce the ship's tendency to "crab" or slide sideways. The tug has direct control over the most maneuverable end of the ship, which increases the tug's ability to rotate this end around the ship's pivot point.

Multiple Tugs Alongside

When multiple tugs are used, the functions of propelling and steering the ship can be singular to or shared by each tug. When two tugs are used alongside, one is the primary power tug and the other the primary steering tug. A common configuration is to have one made up on the hip as the power tug and the other on a headline as the steering tug (fig. 15-4).

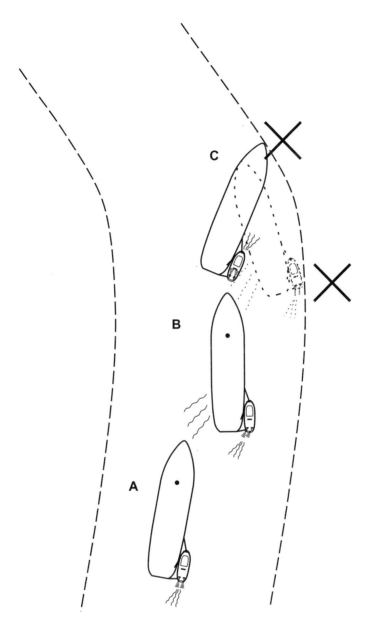

Fig. 15-3. Consequences of a misjudged turn with tug on outboard side of turn. A: This is a critical decision point. At this juncture the pilot still has room to abort the turn. B: The advance and turn rate of the ship have been misjudged. C: The tug backs full to abort the turn but instead counters the turn and slows the ship before the inevitable impact.

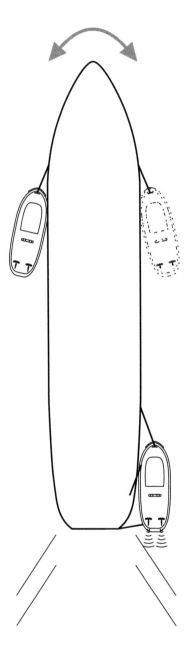

Fig. 15-4. Dead ship move, two tugs alongside. Note that the tug forward can be placed on either bow, but is more effective on the side opposite to the propelling tug aft.

Another means of employing two tugs alongside is to have them made up facing each other on opposite ends of the ship (fig. 15-5). Although originally designed for single screw tugs this arrangement is extremely effective when tractor or ASD tugs are used. This configuration enables precise control over both ends of the ship and allows a fluid transfer of power and steering roles between tugs.

In some ports, three tugs are required as a minimum for shifting large dead ships. In this case, one tug is hipped up aft to provide propulsion. The other tugs are secured forward on either side of the bow with a headline to assist steering the vessel (fig. 15-6A). Very large ships may require four or more tugs to be used in a shift, with two tugs fast alongside aft and two tugs forward, one either side, to steer (fig. 15-6B). The reason for employing so many tugs has to do more with maintaining control over the vessel than the need for the propulsion (this assumes that the tugs are well powered). Tugs so far from the centerline on very wide ships have a limited ability to steer the vessel and frequently are used as if they were machinery components of a twin-screw ship, using one tug's engine ahead and the other's astern at times for the twin-screw effect.

TUGS ON TOWLINES

Two tugs on towlines are often used for shifting dead ships, especially when navigating locks, bridges, and narrow channels. Each tug steers its respective end, while the two tugs balance their propelling forces to advance, retard, or stop the ship. The lead tug is on a towline and the trailing tug is on a towline or headline, depending on its design and propulsion. Tractor and ASD tugs are particularly well suited for this configuration.

The tug at the trailing end of the ship particularly must be alert and check the way of the ship, so that it does not gather enough way to overpower the lead tug. If one of the tugs is a tractor, ASD, or conventional tug with flanking rudders, it is best to have it act as the trailing tug, as it can slow or steer the vessel while backing down (fig. 14-24). If the trailing tug is a conventional tug working stern first, a gobline should be rigged to prevent this tug from being tripped.

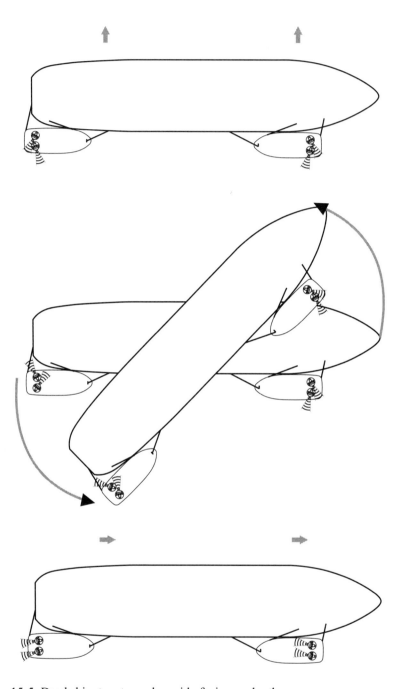

Fig. 15-5. Dead ship, two tugs alongside facing each other.

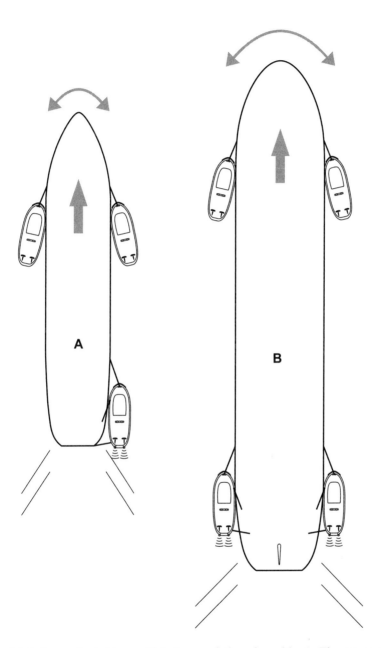

Fig. 15-6. Large dead ship, multiple tugs assisting alongside. A: Three tugs. One power tug alongside, two steering tugs on headline. B: Four tugs. Two power tugs alongside, two steering tugs on headlines.

TUG ALONGSIDE AND LEAD TUG ON TOWLINE

On lengthy shifts, it is sometimes practical to use a tug made fast alongside on the quarter of the ship and a tug ahead of the ship with a towline to the bow. This is a particularly effective way to use two tugs with moderate power to move sizeable ships. When the ship is moving ahead, both tugs are propelling and controlling the vessel. When the tug on the stern stops the ship by backing, the tug on the bow can overcome the ships tendency to swing by pulling to offset this effect (fig. 15-7).

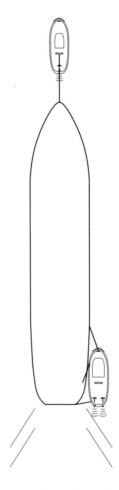

Fig. 15-7. Dead ship, two tugs, one alongside and one on towline.

The size and shape of most ships today make them wholly impractical to handle with just one tug alongside. The extreme flair and overhang may rule out the option of placing a tug alongside at either end of the ship and the sheer size of most ships mandate the use of two or more tugs. Figure 15-8 illustrates some alternative configurations for using multiple tugs to move large, dead ships.

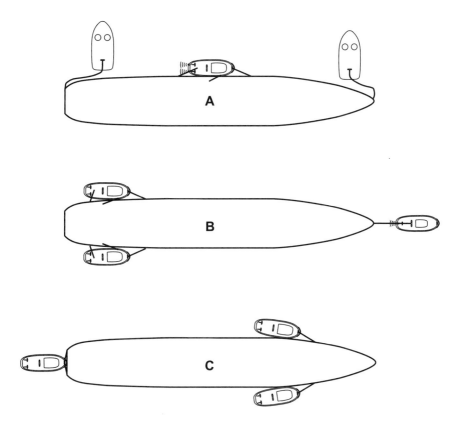

Fig. 15-8. Large dead ship, multiple tugs. Position A: Power tug is positioned midships due to ships overhangs. Two other tractor/ASP tugs push/pull to shift ship laterally. Position B: One tug on towline, two power tugs alongside. Position C: One tug pushing on transom, two steering tugs on headlines.

STUDY QUESTIONS

1. What is the difference between handling a dead ship and assisting a powered vessel?
2. Is one tug ever used to handle a small ship by itself?
3. If only one tug is used and a hard turn must be made, where should the tug be?
4. How may a tug on the outboard side of turn reach a point of no return?
5. Does a ship propelled by a tug on the hip have a tendency to skid sideways?
6. When a larger ship is shifted, are more tugs required?
7. Are vessels ever handled stern first?
8. Describe several hookups for handling dead ships.
9. Why is the use of a tug on a towline forward and a tug hipped up aft an effective way to use tugs?
10. How are tugs fast aft on beamy vessels used most effectively for steering?

CHAPTER SIXTEEN

TUGS AND ANCHORS

The anchor is one of the oldest and most basic tools in shiphandling. When used as an accessory to shiphandling, the anchor often is referred to as a "poor man's tugboat." This is an interesting but not altogether accurate analogy. Using an anchor sometimes allows the shiphandler to dispense with the services of a tug and at other times may be preferable to a tug.

However, anchors function differently from tugs. They are essentially passive devices. They can neither push nor pull like a tug, unless they are heaved in after they have been dropped. When the ship is docking or maneuvering, the anchor has three principal functions:

- restrain but not stop the ship
- fix or hold the anchored end of the ship
- provide a deadman to heave against

For example, the anchor can be used to stabilize the light bow of the ship on a windy day if enough chain is slacked to allow the anchor to drag along the bottom. If the bow continues to blow off, then more chain is slacked—carefully—until the ship responds. When the proper amount of chain is slacked out, the resistance of the anchor dragging through material on the bottom helps a ship maintain steerage at very slow speeds, with the engine working ahead. It may not even have to back its engine to stop when docking, since the resistance of the anchor may be sufficient to check its way when the engine is stopped. The amount of chain let out is critical and depends on the depth of the water and nature of the bottom. A shot to a shot-and-a-half at the water's edge is a good benchmark.

The only time an anchor can be used to pull is when it is used as a brake to check a vessel's way and as a deadman (or a kedge) to heave against. It cannot push at all. When it is used as a deadman, the bow (or stern) of the vessel can be maneuvered by heaving while a tug pushes or pulls the other end of the ship. For example, it can be very handy for undocking a ship (fig. 16-1).

When employed passively in conjunction with a tug, the anchor may be used to:

- restrain the vessel's motion (ahead or astern)
- stabilize a vessel's direction of movement
- hold (subject to the amount of chain used) the position of the bow (or the stern—if a stern anchor is used) while a tug maneuvers the other end of the ship

There are a number of maneuvers somewhat like the Mediterranean moor where the vessel is turned through 180°. The tug is used to assist the vessel through the turn, while the anchor stabilizes the bow (figs. 16-1, 16-2).

The anchor can be used to provide directional stability to a vessel being towed stern first by a tug or when that vessel is backing with a tug made up on the quarter to steer the stern (fig. 16-3).

Some ships are fitted with stern anchors. They can be useful, especially when the vessel is required to anchor in confined areas or docking and undocking in fair current situations (fig. 16-4).

When tugs and anchors are used for shiphandling, the pilot should exercise the same degree of control over the anchor that he does over the tug. The mate on the bow must understand the pilot's orders to ensure that the anchor is not dropped too soon or too late. If too much chain is slacked, it may not be possible for the anchor to be heaved back aboard the ship once it is berthed (if this is necessary). Conversely, if too little chain is slacked, it may not have the desired effect and the vessel will not respond as expected.

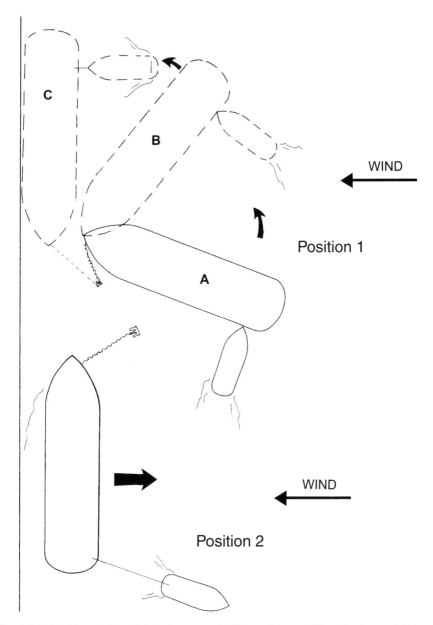

Fig. 16-1. Position 1: Vessel is being turned. The anchor stabilizes the bow while the tug pushes the stern around. The anchor will prevent the bow from landing heavily on the dock. Position 2: Ship uses the anchor to heave the bow off the dock while the tug tows the stern clear.

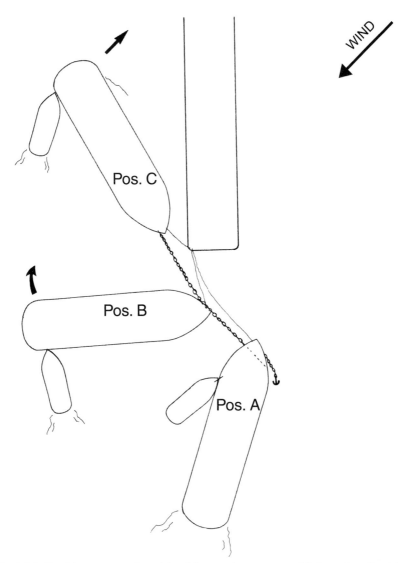

Fig. 16-2. Docking, stern to/lee side of finger pier, one tug, ship's anchor. Position A: Tug is placed forward until the anchor is dropped and a headline passed to the end of the dock. Position B: Tug has shifted to the opposite quarter and will turn the ship. Position C: Tug is breasting the ship to the dock and easily pushing it astern. The wind is holding the bow of the vessel off the dock. The anchor chain must be slacked judiciously and the headline tended on the capstan. Note that these are common techniques employed by trampers in remote ports where tug availability may be limited.

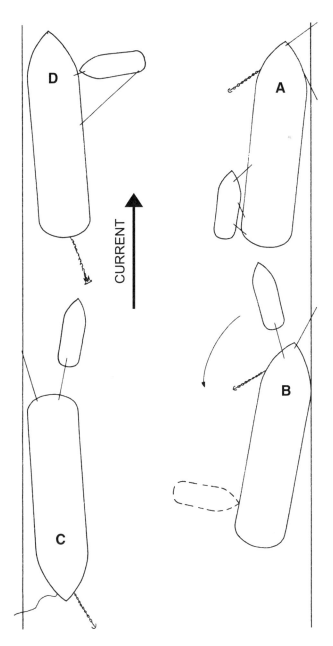

Fig. 16-3. Tugs being used in conjunction with anchors to berth ships. Tug handling vessel B may have to shift to breast the stern of the ship alongside the dock, depending upon the direction of the wind. Tug handling vessel C is towing the vessel stern first into its berth.

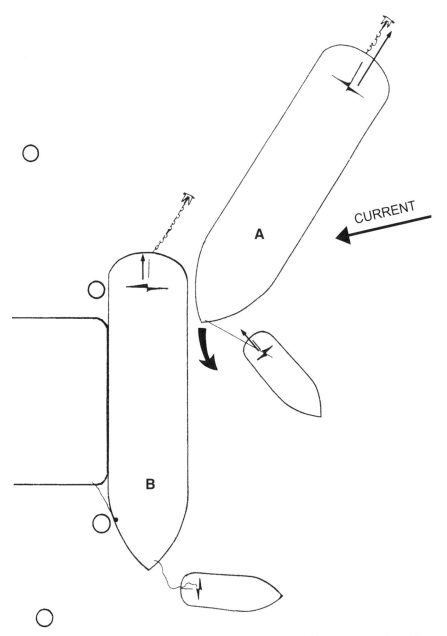

Fig. 16-4. Docking with fair current on dock, one tug, ship's stern anchor. Note that the pilot should exercise caution when stern anchors are used and make sure that the chain always leads astern. If the vessel overrides the chain when backing or maneuvering, the rudders and propellers could be damaged.

When an anchor has been used as a deadman, the chain must be heaved in when the vessel is unberthing. If it is heaved too fast, the anchor can cause the vessel's bow to swing more than is desirable, making the vessel difficult to control.

Using tugs and anchors requires teamwork and coordination between the pilot, ship's crew, and tug operator. When executed smartly, shiphandling with tugs and anchors represents a pinnacle of professional seamanship.

STUDY QUESTIONS

1. Why is an anchor called a poor man's tugboat?
2. How does using an anchor differ from using a tug?
3. What are the three functions of an anchor when a ship is maneuvering?
4. How is the anchor used when tugs are also employed?
5. How are stern anchors used?
6. What is the danger when heaving a stern anchor?

CHAPTER SEVENTEEN

USING TUGS IN EXPOSED WATERS

In the mid-twentieth century the use of tugs to assist ships in exposed waters was limited to salvage or rescue operations. Farley Mowatt gives a vivid account of this type of tug work in *Grey Seas Under*, a book that follows the journeys of the *Foundation Franklin*, a salvage tug working the North Atlantic waters off Nova Scotia.

While shipwork with tugs in exposed waters may be less adventurous than the doings of the *Foundation Franklin*, tugs today routinely assist ships in open roadsteads. This owes both to the growing practice of using tugs to escort tankers and other large product carriers from sea, and to the escalation of ship sizes that have outgrown the berths and waterways of many ports. Some ships are of such size that only a few harbors are large enough to accommodate them. Consequently, they often must load and discharge their cargos at moorings and berths located in open roadsteads. Ships also may require a tug assist in open waters when they are approaching harbors with narrow channels, particularly if the current is setting heavily across the channel. Even in exposed waters tugs must perform their basic function of pulling and pushing to assist ships to safely maneuver, dock and undock.

TUG EQUIPMENT

Fendering

When working in exposed waters, both ship and tug risk damage from heavy contact due to sea conditions that do not exist in ordinary shipwork. The tug should be especially well fendered for this type of service. Fenders must be ample and well secured to avoid coming adrift when the tug pitches heavily or pushes against the ship. Fenders should protect tug and ship but allow the tug to slide easily up and down the ship's side when working in a swell. Although a tug fender's capacity to grip or stick to the side of the ship may be desirable in protected waters, it can be a detriment offshore.

Tug and fender designers employ various methods to lower the friction component of the tug's fenders. The texture of the fenders surface may be smooth, the fender may be constructed with an outer layer of low friction material, and water may be used as a lubricant. Some tugs have heavy pneumatic wheel-type fenders that roll up and down the ship's side as the tug pitches (fig. 17-1). These fendering innovations are a necessary and welcome improvement over conventional fenders for working tugs in exposed waters.

Lines and Winches

While fendering is important, lines and winches also must be considered. Working lines should be larger than normal size. They should be given a longer lead to prevent them from parting, since they come under heavy strain during the rise and fall of the tug in the swells. Strength is not the only relevant factor in choosing an appropriate line size. Abrasion resistance also should be considered. In a seaway the tug's lines are subject to extreme loads and abrasion where they pass through chocks on both the tug and ship. Although a smaller line may have a sufficient strength rating, there are reasons why an oversized line may be more appropriate in exposed waters. The larger diameter of the line also increases the area of the line's bearing surface where it passes through the chock and increases accommodation for line abrasion.

Winch technology offers another means of managing the dynamic loading of lines in a seaway. Automatic render and recovery winches pay out and pull in line as line-loads fluctuate with the tug's motion and bollard pull in a swell.

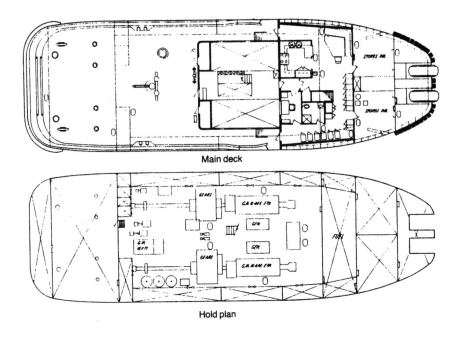

Main deck

Hold plan

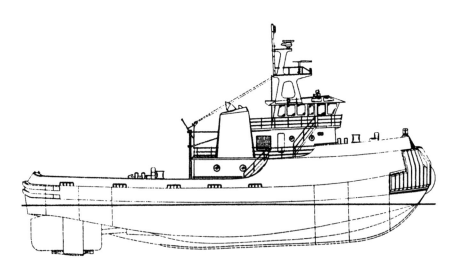

Fig. 17-1. Twin-screw tug of 5,700 horsepower designed for docking VLCCs and ULCCs in exposed areas.

OPERATIONAL CONSIDERATIONS

Deep-loaded ships usually are fairly stable platforms, and tugs often can make up safely in their lee in fairly rough seas. Ships in ballast often roll quite a bit, and special care must be taken as the tug comes alongside, since the vessels can roll against each other. Contact between the ship and tug in a heavy seaway can cause considerable damage, even to well-fendered tugs.

Tugs respond more slowly to their engine and rudders in a seaway than they do in quiet water. Regardless of his skill, the tug operator is unable to maneuver with the same finesse in rough water as he would in calm conditions. The pilot or docking master must anticipate and take into account this negative effect of a seaway on the tug's maneuverability.

Towline work particularly is suitable for shiphandling in exposed waters. It minimizes or eliminates contact between the ship and the tug. The tug operator can increase towline length to moderate the effect of the swells on the strain of the towline.

When doing towline work in exposed waters operators should be aware particularly of deck edge immersion. The heeling angles acceptable in protected waters may be excessive and hazardous in exposed waters. It is immaterial whether the deck edge is immersed due to a heeling angle or the open sea and swells heaping water on the deck. Once solid water, from any source, consistently submerges the deck while the tug is on the towline, risk of foundering and flooding escalate exponentially. Newer tugs dedicated to escort service have been designed to minimize this risk (fig. 17-2).

SINGLE POINT MOORING

A common use of tugs in exposed waters is to assist a ship making fast to a single-point mooring (SPM). Many of these moorings are located in unsheltered areas and subject to a variety of wind and sea conditions. The large vessel that normally loads or discharges at this facility must make an extended approach to the mooring buoy. This is a delicate balancing act between the forces of wind and current and the ship's often-limited maneuverability at slow speed. And it is often a protracted process since it takes a while for the ships headway to dissipate in order to make the final approach with bare steerageway. Unless the current is running strongly, bare steerageway is lost as soon as the vessel backs its engine to stop. At this juncture the vessel must be close enough to the buoy and remain in this position long enough for the mooring gear (usually chain or cable) to be passed aboard and secured.

Where wind is not a factor, a ship with a right-hand propeller should approach the buoy from down current with the buoy dead ahead or fine to port. The pilot is anticipating the effect of the propeller's torque. When the ship backs, it will clear the buoy but still be close enough to pass the lines. This is a ticklish maneuver, but not necessarily dangerous unless the SPM is a structure. However, much time can be lost if the shiphandler misjudges the situation or if the ship handles poorly. A tug can provide some welcome assistance by helping the vessel to steer—particularly during the final approach, and especially keeping the ship in shape when it backs. This can be quite helpful when the wind is at odds with the current, and if the ship has a tendency to weathervane (to head or to back into wind) as soon as the engine is rung astern.

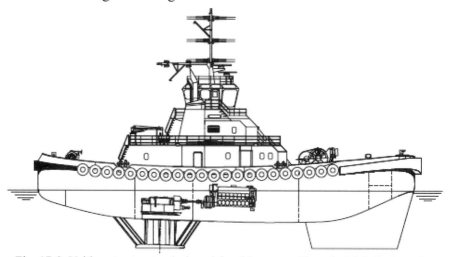

Fig. 17-2. Voith water tractor designed for ship escort. Note the high freeboard for towline work in exposed waters. (*Courtesy of Voith Turbo Schneider Propulsion GmbH & Co.*)

A tug is best employed alongside forward and clear of the chocks where the mooring gear is to be passed. A tug on a towline might interfere with the mooring operation (fig. 17-3). Since the ship is moving very slowly and must stop to pick up the mooring, a tug of relatively moderate power can be useful in this situation. Many SPMs are unattended by tugs but, if tugs are available, they can be used to avoid unnecessary delays and to eliminate the uncertainties that are common when large ships are handled at slow speeds.

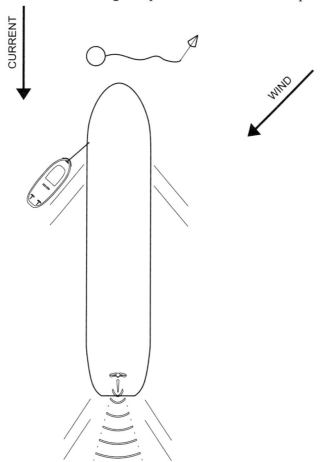

Fig. 17-3. With a large tanker approaching SPM, the tug may be required to back or push to keep the vessel in shape when it backs its engine to stop.

STUDY QUESTIONS

1. What additional considerations must be given to a tug assisting ships in exposed waters?
2. What is a roller fender and is it suitable for this type of shipwork?
3. In addition to using roller fenders, what other precautions should be taken?
4. Does a tug handle the same in a seaway as it does in protected waters?
5. How does a tug assist a vessel mooring to an SPM?

CHAPTER EIGHTEEN

SALVAGE AND EMERGENCY SITUATIONS

Nearly all marine casualties occur in pilotage waters. Since tugs are usually available, they often are called upon to assist vessels in salvage and emergency situations. The terms salvage and emergency are not synonymous or mutually exclusive. The distinction hinges on legal technicalities that are not material to this text. The two subjects are dealt with separately in this chapter.

SALVAGE SITUATIONS

Exclusive of acts of war or fire, marine salvage normally falls into one of three categories: sinking, grounding or stranding, and rescue towing. Marine salvage can be a very complicated business, and much of it is beyond the scope of this book. Our interest is confined only to that category most likely to confront the pilot or docking master—grounding.

Vessel Aground

Almost inevitably, every shiphandler has to deal with grounding on some occasion. If he puts the vessel aground, it is likely that he also will be engaged in refloating it. If the vessel is too hard aground or stranded, the services of a professional salvor may be required. In any event, the shiphandler should know something about both the forces to be overcome to refloat the vessel and the remedies. Once a vessel is aground, four different factors can affect it and the efforts to free it:

1. High winds can have a considerable effect, especially on light-draft vessels with a lot of exposed freeboard. They can cast a vessel ashore and keep it there. Conversely, if the wind shifts, it can free the vessel.

2. Strong currents have their greatest effect on deep-draft ships and, like heavy winds, can beach a vessel and work it even farther ashore. They also can cause a heavy buildup of sand or mud around a vessel (usually in proximity to the up-current side) that can complicate refloating it.

3. Heavy seas can put a vessel ashore and force it harder aground. They can also frustrate rescue efforts and cause serious damage. Paradoxically, heavy seas can also work to the salvor's advantage. They may help break the suction of sand or mud bottom, grind down a coral bank, or (if the vessel is pitching) "pump" clear the loose material beneath a vessel and help free it.

4. Ground effect is the amount of the vessel's lost displacement and the actual weight of the vessel borne by the material that it rests upon. This can be determined, if sea conditions permit, by comparing the vessel's draft before and after grounding. The vessel's immersion scale will provide the necessary information to calculate the amount of ground effect in tons.

The friction or resistance that must be overcome to refloat the vessel is a coefficient of this weight (ground effect) and ranges from 30 percent to 150 percent or more, depending on the nature of the bottom. Salvors calculate this percentage at 30 percent to 40 percent for soft sand or mud, about 50 percent for hard sand or gravel bottom, 50 percent to 100 percent for coral bottom, and 80 percent to 90 percent, up to 150 percent, for rocky bottom.

Vessels can be refloated in several different ways. If a vessel is grounded at low tide and the tidal range is sufficient, it might float off on the succeeding high tide. But if the ship went aground at or near high tide, it might have to wait until the next spring tide to be refloated—and then only if there was a sufficient difference in the tidal range.

A vessel might be able to lighten up enough by discharging ballast or cargo or by jettisoning enough cargo (the ultimate sacrifice) to refloat itself. But if these methods fail or are not feasible, the remaining alternatives involve using beach gear or assistance from tugs.

Salvors employ any and all of these methods—singularly or collectively if necessary—to refloat a ship. Each method has its pluses and minuses. Waiting for the tide is cheapest if the ship does not have to wait too long for a high enough tide. But even in this case, it might be wise to have a tug standing by to avoid having the vessel broach when it is free of the bottom.

Discharging ballast, if prudent or practical, is the next simplest method. The salvor should consider the effect this will have on the vessel's stability, since the ship will already have suffered a loss of GM (metacentric height) because of the shift in the center of buoyancy.

Discharging cargo is an acceptable method of lightening ship, but sea conditions may not permit the necessary lighters and barges to lie alongside. Furthermore, usually only break-bulk freighters and tankers are likely to have the necessary gear aboard to carry out the discharge unassisted. Jettisoning cargo is a last resort. In the case of a tanker loaded with crude, it is not a viable option.

Beach gear or tugs often are employed in conjunction with other efforts and are almost certain to be used when other alternatives have failed. Beach gear was the traditional method of refloating stranded ships. This consists of heavy wire rope falls attached to a strong cable that is secured to an anchor and several shots of chain for weight and set offshore of the ship. Sometimes the gear is laid out on the deck of a barge fitted with winches and attached to the ship by heavy cables. In other instances the ship may use one of its own anchors attached to a cable and rig its own tackle gear on deck. The falls in this case have sometimes been contrived from the vessel's own cargo gear. The average set of beach gear can develop forty to sixty tons of pull. Often several sets of gear are used to provide sufficient power to pull the vessel off. Beach gear rigged on barges is still used, but many modern salvage tugs are fitted with powerful winches and heavy anchors (usually Eell-type anchors) that can exert up to one hundred tons of pull on the anchor cable. The salvage vessels usually set out their anchors and then pass their tow cable to the stranded ship. This arrangement has the same function as conventional beach gear but is easier to set up.

Tugs usually are the first vessels called to assist a ship that has grounded, since they are available on short notice. Usually one tug is insufficient to refloat a large ship that is hard aground; different-sized and powered tugs may be necessary. Smaller tugs may be required to transfer personnel, equipment and lines, and handle or assist barges with beach gear. Larger, high-horsepower tugs may be necessary to exert enough bollard pull to free the ship.

High-horsepower tugs are now available that can generate bollard pull in excess of 200,000 pounds. These powerful tugs can provide a competitive alternative to standard salvage operations using beach gear. One reason is that beach gear generates a static pull in the direction that the anchor is placed. The tugs, on the other hand, can move about and pull from different directions. When this is done, tugs can twist the vessel from side to side, which helps break the grip of suction from the bottom. If the vessel is aground at one end, or in one area of its bottom, the overhanging portion of the ship becomes, in effect, a giant lever. The tug can pull on the end of this lever which can greatly increase the tug's effectiveness.

Certain elementary steps should be taken as soon as possible after a vessel goes aground:

1. Determine if the vessel is holed.
2. Check the vessel's draft to determine how much it is aground.
3. Take soundings around the vessel to determine where it is aground.
4. Note the state of the tide when the vessel grounded, and determine whether the range is increasing or decreasing.

The reasons for these measures are self-evident. If the vessel is holed, for example, and the ship is making more water than its pumps can control, it would be prudent to leave the vessel aground until the leak is repaired or more pumps are available. The difference in draft before and after grounding will give some idea of the amount of ground effect, and soundings around the perimeter of the ship will locate the portion of the ship that is aground, and perhaps indicate the area of the vessel's bottom where damage is likely to have occurred. This also may be a good indicator of how the tugs can be employed most efficiently. Vessels aground all along the bottom in soft sand or mud usually come off best in the direction opposite to the one in which they went aground. Of course, twisting the vessel helps break the suction of the bottom against the hull, which may be a large component of the resistance necessary to overcome.

Sometimes it helps to spot the tugs so that their wheel wash helps dredge away material from the side of the vessel. This is best accomplished by making the tugs fast on a short towline at various stations along the ship's side and moving them progressively from one position to another if this action is effective.

If the vessel is aground on one side only (which sometimes occurs when a pilot overshoots a bend), it may be possible to have the tugs push against the inshore side of the ship, providing there is enough water there. Unless the bottom is quite soft, it is probably best to have the tugs concentrate their effort near the stern. This permits the vessel to back clear as it comes off without damaging the stern gear (rudder and propeller.) (See fig. 18-1.)

Vessels that ground on a lump or bank and are only aground in one place (fore, aft, or midship) usually respond best if they can be swung from side to side. On sand or mud bottom this tends to flatten the lump, and on coral or limestone bottom it has a tendency to grind down the material until the vessel comes free. Vessels aground on steep-to, firm mud banks, or rounded cobble stones can come off with very little warning. Both the ship's and tug's crews should be warned to anticipate this possibility.

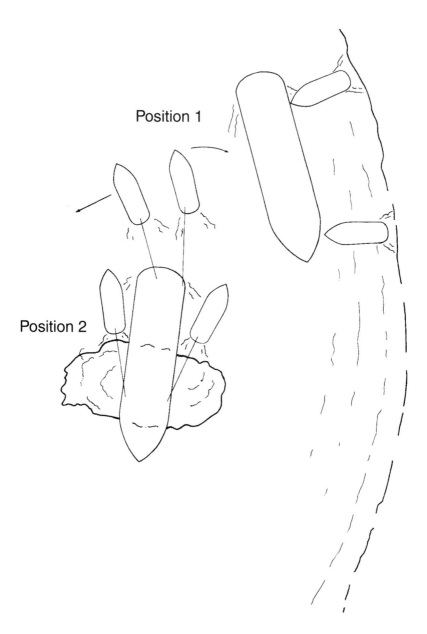

Fig. 18-1. Position 1. Vessel has grounded on the bend. Tugs pushing on inboard side probably are most effective in refloating vessel. Position 2. Vessel may find it advantageous to have tugs pulling from alongside to dredge material away from shipside. Tugs pulling on stern should swing from side to side to twist the vessel, as this helps break the suction.

The groundings that a shiphandler is most likely to encounter usually occur in protected waters. This makes salvage efforts easier and safer. But occasionally ships go ashore in areas exposed to heavy seas or wave action, often just outside the breakwater or harbor entrance.

If a vessel does ground in an exposed area and a tug can come alongside safely, it may be wise to have it spot the ship's anchors to weather to prevent it working farther ashore. The anchors are not likely to be much help refloating the ship if the vessel drove ashore bow first since the lead of the chain will be foul in the hawse, unless it is a stern anchor. Anchors can be detached from their chains and connected to cables that can be heaved with falls laid out on deck of the ship like beach gear (fig. 18-2).

If it is not possible to set the anchors, the vessel should ballast down if the sea makes up to avoid going harder aground. Even if only one small tug arrives, it should put up a line and take a strain, until more help arrives. This may prevent the ship from "climbing the beach" or broaching. If a vessel broaches on a rocky shore more hull damage can occur. Sand beaches have a tendency to swallow the ship. On a coral or limestone shore the vessel has a tendency to dig itself in and be much harder to refloat.

Tugs should be ready to make their maximum effort at the peak of the flood tide. Fuel and ballast should be shifted or discharged if it is helpful to lighten or trim the ship beforehand. The pilot or salvage master should be in constant radio contact with the tugs when they are working. If the vessel comes off suddenly, and if communication is not maintained, the results can be disastrous. This happened once with nine tugs working on towlines. The ship came off as though propelled by a giant slingshot. Fortunately, only one tug was seriously endangered; its hawser parted, saving it from a probable capsize.

No two salvages are exactly alike, but certain fundamental rules apply:

1. If possible, refloat the vessel at high tide.

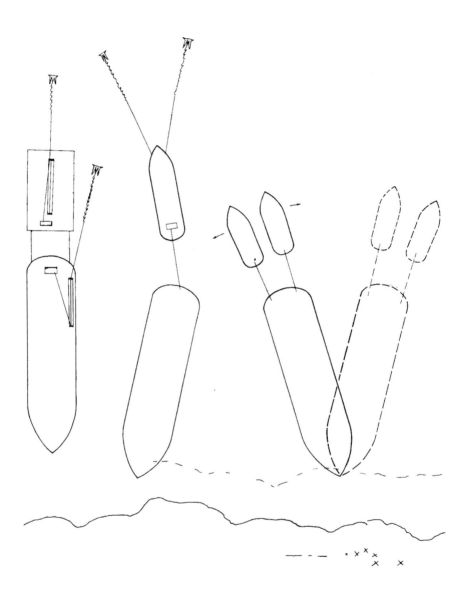

Fig. 18-2. Different aspects of refloating a vessel by traction are shown: beach gear rigged on a barge; beach gear rigged on the vessel; a salvage vessel with the anchor cables led to winches; and tugs. The tugs can pull from different directions to twist the vessel.

2. If the bottom is soft and the vessel is aground fore and aft, tugs should be placed where their wheel wash will dredge away material.

3. If tugs are working close together on towlines, they should adjust their hawsers to the same length. If the tugs come together they are usually well fendered and no damage is likely to occur. However, if one tug strikes another's hawser it is likely to part and cause severe damage or injury.

4. If the vessel starts to move at all, even from side to side, the tugs should maneuver to sustain the motion.

5. Even if a vessel can refloat itself with its own engine, it may require a tug to assist in steering until it reaches deeper water.

6. A successful salvage is a result of a coordinated effort. Salvage masters and tug operators should stay alert and maintain contact during the entire time they are working.

USING TUGS IN EMERGENCIES

Most emergencies affecting vessels in pilotage waters fall into one or more of five categories:

1. Fire
2. Holed or sinking
3. Stranded (discussed above)
4. Loss of power
5. Loss of steering

Tugs are suitable for assisting in most of these situations since they are powerful and maneuverable. If they are fitted with fire-fighting gear, as many are, their capabilities in an emergency are enhanced.

Fire

Fire may not be the most common emergency but is potentially the most dangerous. It often is essential, especially in the case of tankers, to remove the tankers from their berths. This is likely to be the tug's first priority. Ships normally have fire wires hung over the side, fore and aft, for tugs to pick up. Before approaching the ship, the tug should have its towing gear ranged (towline or cable, shackles, etc.), fire hoses led out and tended, and the crew dressed in oilskins or fire-fighting gear.

The tug will back up to the ship to take it in tow astern. Fire hoses can be played upon the ship to reduce the heat and protect the crew while they are securing the towline to the fire warp. Other vessels can help by directing their fire monitors or hoses in such a fashion that they will cool the area. The tug should slack 200 to 250 feet of hawser or tow cable before taking an easy strain in order to avoid parting the fire wire (fig. 18-3).

If the tugs are secured at both bow and stern of the ship, they can control the vessel fairly well. However, if only one tug is used, it probably is best secured forward. The ship will be easier to control in this configuration, unless the vessel is trimmed heavily by the bow.

Fire wires are, at times, hung on the offshore side of the ship's tug or bow. If a tug picks up one of these fire wires and begins towing, the ship tends to sheer away from the side that the tug is secured to. This tendency normally is less pronounced if the tug is made up to the bow wire rather than the stern. In either case, if a tug can get alongside at the opposite end of the vessel and push against the same side that the towing tug is on, it can assist in maintaining control over the ship (fig. 18-4).

Once the vessel is clear of the dock, it is often convenient to ground it. This should be done in an area where the bottom is flat and soft so that the ship will not be unnecessarily damaged when the tide drops. If the tug is required to assist at fire fighting, it will usually be directed by professional fire fighters.

Vessel Sinking or Holed

If the ship is sinking or holed, the tug needs to assist in putting it aground as quickly as possible.

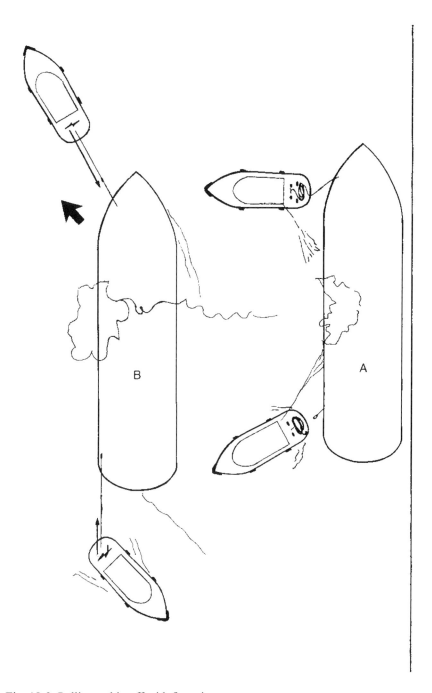

Fig. 18-3. Pulling a ship off with fire wires.

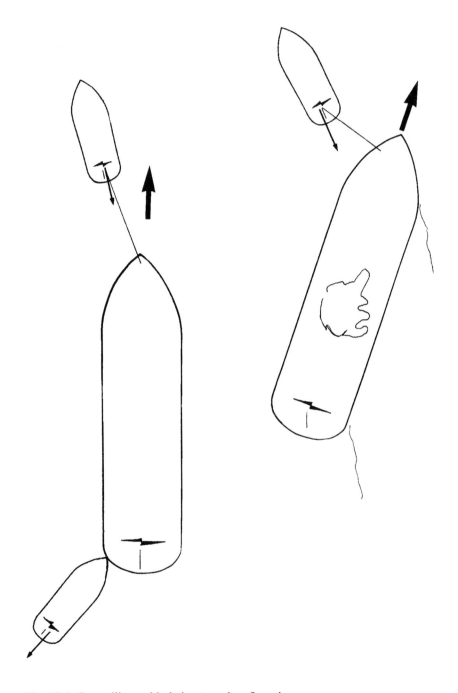

Fig. 18-4. Controlling a ship being towed on fire wires.

If a ship has been holed while at a dock, tugs have been used to breast a vessel against the dock to keep it from capsizing or sliding into deeper water.

If a tank vessel is holed and leaking flammable products, the tug should stay clear of this area, since the atmosphere may be explosive or toxic. The fumes could cause a galley or an engine room explosion. The tug may be able to approach the vessel only from the weather side, but even this might be hazardous (fig. 18-5).

Loss of Power

This is a dead ship situation. By using a towline, a small tug can handle a large ship that would be impossible to handle on the hip. If the tug is made up to the ship's bow and is having trouble controlling the ship, the tug can take the vessel in tow stern first while the ship drags one of its anchors for directional stability.

Loss of Steering

A tug can readily assist a vessel with loss of steering to steer. It can work nicely on a towline ahead of the vessel especially if the vessel is light forward. The vessel should steam ahead slowly to avoid tripping the tug.

The tug also can steer the vessel by making up alongside forward, where it can control the ship at moderate speeds by pushing or backing. It can also hip up on the quarter of the vessel and employ the tug's engine in opposition to the ship's engine (ahead and astern) for a twin-screw effect. The tug's rudder can also be helpful in this instance.

The tug also can steer the vessel by pushing on either side of the stern as needed, or by using direct or in-direct towing methods to sheer the stern one way or the other.

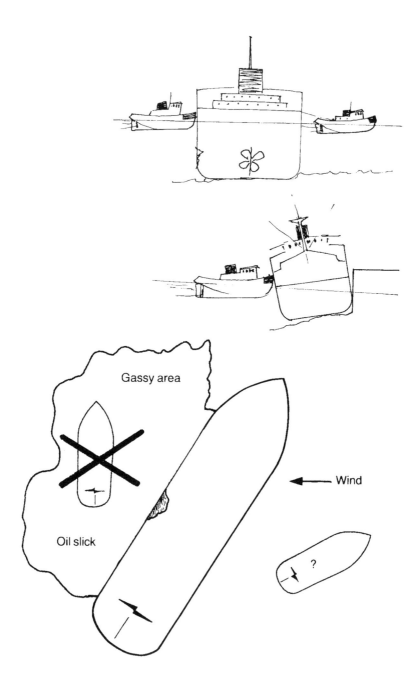

Fig. 18-5. Holed or sinking vessels.

Some Priorities

Tug masters should be aware of the order of their responsibilities. Their first priority is the safety and welfare of their crews. Their next concern is for their own vessel. Their final consideration is the vessel they are assisting. They might be justified in risking their crews to save lives, but they must weigh the risk carefully before exposing their personnel to undue hazard to save property. Those who employ a tug master's services also should be aware of these responsibilities.

STUDY QUESTIONS

1. When a vessel goes aground what forces must be overcome to refloat it?
2. What procedures are used to refloat a vessel?
3. Which is the least desirable method?
4. What are the first steps to be taken after a vessel goes aground?
5. If a vessel is holed, should it be refloated?
6. What is one of the effects of current?
7. What effects can heavy seas have on a vessel aground?
8. What should a vessel do to avoid being washed farther aground?
9. If the vessel can refloat itself, why is it sometimes prudent to use a tug?
10. How may a small tug assist a ship that is aground?
11. What is beach gear?
12. Are tugs sometimes more effective than beach gear?
13. To what type of emergencies might a tug have to respond?
14. How should a tug take a ship in tow that is on fire?
15. Can one tug control the ship?
16. How can a tug assist a ship that has a power failure?
17. How can a tug assist a vessel that has lost its steering?
18. If a tank vessel is holed and the area is gassy, what should the tug do?
19. List a tug captain's responsibilities in order of importance in an emergency.

CHAPTER NINETEEN

SHIP ESCORT

Using tugs to escort ships from sea is not a new business. Since their introduction in the early nineteenth century, tugs have been shepherding vessels from open roadsteads into canals, rivers, and other restricted waterways.

However, since the mid 1970s, a specialized form of ship escort has evolved, concerned primarily with the movement of oil tankers. The enormous environmental and economic consequences of a grounded or sunk tanker are well documented. International, national, and regional groups have sought and demanded risk management tools that would eliminate or minimize the probability of catastrophic tanker incidents. Tugboats are one of those tools, but are not a panacea.

Many human, environmental, and mechanical factors contribute to tanker incidents, not all of which can be managed by the presence of a tugboat. Nevertheless, in some circumstances a tug of the proper design and horsepower literally can alter the course of a calamity by being present in the event of a tanker's steering or propulsion failure.

Detailed analysis and planning are required to ensure that a tug of sufficient capability is in place at the time it may be needed. It also requires a ship pilot and tug operator who can use the tug to its maximum capacity for the intended purpose.

The primary purpose of tugs in ship escort is to provide:

- emergency steering
- emergency stopping

In the event of a loss or malfunction of the ship's steering or propulsion, emergency steering and stopping are fulfilled through employment of three primary ship/tug response modes (fig. 19-1):

- retard
- assist
- oppose

The objective in the Retard Maneuver is to take the way off the tanker as quickly as possible without concern for the tanker's course. The intent of the Assist Maneuver is to enhance the effect of the ship's rudder and make the ship turn as tight as possible. The goal of the Oppose Maneuver is to oppose the turning force

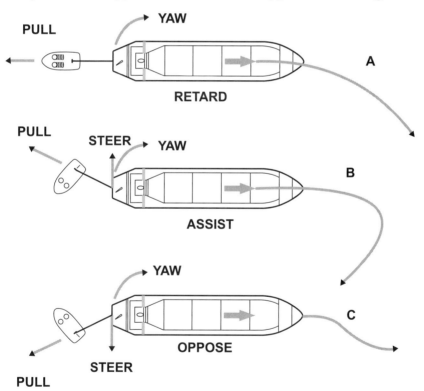

Fig. 19-1. Ship/tug emergency response modes. Position A: Retard by pulling astern. Position B: Assist the ship's rudder in a turn. Position C: Oppose the ship's rudder in a turn.

of the ship's rudder and slow the ship's turn rate or hold the ship close to its original heading. These three modes are employed in combination or individually depending on circumstances.

In the event of either an engine or steering failure, the ship likely will veer to one side or the other. In that case, the escort tug is called on to both retard and steer the ship. The success or failure of the emergency response maneuvers depends on many environmental, human, and equipment factors. Of those, tug design, escort tug position, ship speed, and time delays are at the forefront.

TUG DESIGN FACTORS

Tugs called on to escort ships may be general- or special-purpose escort tugs. It is important for both pilot and tug operator to recognize the performance differences between the two. The general-purpose tug may meet the bollard pull and free running speed requirements of an escort plan, but its performance may not be at the level of a special-purpose escort tug (fig. 19-2).

Special-purpose escort tugs are designed to escort ships at relatively high speeds over long distances. New-built escort tugs are exclusively omni-directional (VSP or ASD) and have the stability, freeboard, horsepower, and deck equipment to apply steering and braking forces at high speeds.

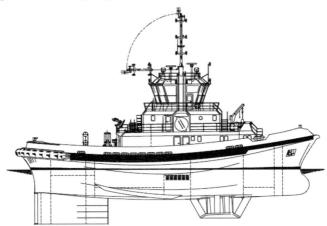

Fig.19-2.Voith-Schneider ship escort tractor tug. *(Courtesy of Østensjø Rederi AS & Voith Turbo Schneider Propulsion GmbH & Co.)*

The choice of a tug's response position must account for differences in tug design. To be effective, a conventional tug must be in a position to push, except at very low escort speeds (three to five knots). In general, effective application of a conventional tug's bollard pull is limited to ship speeds of seven knots or less. At speeds greater than three to four knots the push of a conventional tug loses effectiveness in applying steering forces.

A harbor tractor tug may have the maneuverability to apply indirect towing forces but may not have the skeg design or stability to safely do this at the high speeds (eight to twelve knots) of a special-purpose escort tug.

ESCORT TUG POSITIONS

Positions for tugs engaged in escort work fall into two categories:

- escort position
- response position

Escort Positions

Escort tugs run with the ship either untethered (no line up) or tethered (line up). These are referred respectively as passive and active escorting. *Passive* escorting requires the tug to keep pace with the ship in a position abeam and slightly forward or aft of the tanker. The tug must be in proximity to the tanker and be in a position that gives it the flexibility to maneuver to the optimum response position in a minimum amount of time. Washington State, a pioneer in developing ship escort plans, uses a standard passive position of two tugs abeam and slightly aft of the bridge wing (fig. 19-3).

The most common *active* escort practice is to have a tractor or ASD tug with its towline fast to the ship's stern. Tow line lengths of 300 to 450 feet are common, although some situations may call for shorter lengths. At slower speeds a tug also may actively escort by putting a towline up to the bow of the ship.

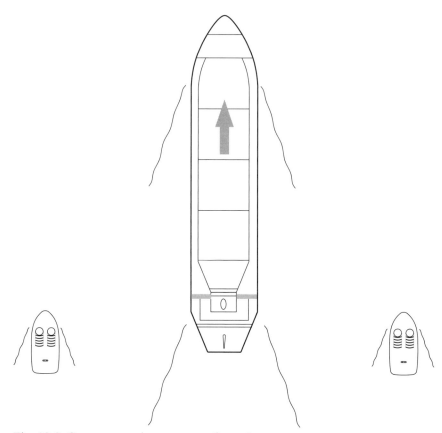

Fig. 19-3. Common passive escort configuration.

Response Positions

The most effective response maneuver at the time of the incident is determined by several on-scene factors:

- relative size of the tanker and tug
- design limits of the tug
- ship speed at the time of failure
- magnitude of the failure
- time delays in recognition of the failure and response
- time delays in applying force to the ship
- on scene wind and wave conditions

Tugs can apply steering and braking forces from five basic response positions (fig. 19-4). Each has inherent advantages and disadvantages that, along with the tug's design capability, determine the position's degree of effectiveness.

TUG IN FRONT OF SHIP'S BOW ON TOWLINE

The tug towing on a line has the advantage of being able to apply steering forces equally to both port and starboard (fig. 19-4A). However, the tug is working on a relatively short assist lever (distance from the ship's pivot point) and its towline force applies both a forward and steering component. Tugs in this position are most effective at speeds of six knots or less.

TUG ON THE SHIP'S FORWARD SHOULDER

This is the least effective of the five positions due to several reasons (fig. 19-4B). First, it is close to the ship's pivot point so the tug has poor leverage to turn the ship. Second, at higher escort speeds, the tug can apply force only by pushing. This tends to increase the ship's speed with little steering effect. Third, when pushing, the tug must be in contact with the ship's hull. The on-scene sea conditions may be too rough to attain this position without damage to the ship or tug.

Fourth, a conventional tug can put up a headline and back but only if the ship has slowed to speeds of 5 knots or less.

Fifth, if a tug is backing, the towline force has both a braking and steering component. The ship tends to turn toward the side the tug is on. Two tugs, one on each side of the ship, are required to apply a braking force without turning the ship.

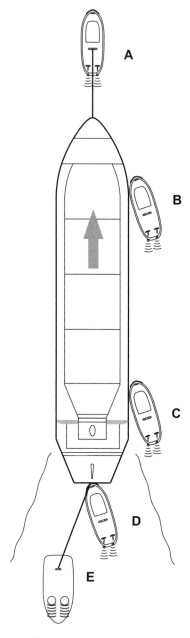

Fig. 19-4. Tug response positions in ship escort service. A: Tug in front of ship's bow on towline. B: Tug on ship's forward shoulder. C: Tug on ship's aft quarter. D: Tug on ship's transom. E: Tug tethered to ship's stern.

TUG ON THE SHIP'S QUARTER.

A tug in this position has many of the same disadvantages as a tug on ship's forward shoulder (fig. 19-4C). When pushing, the effect is to apply both a steering force and add to the ship's forward motion. The on-scene weather must be conducive to the tug pushing on the ship without damage. Ship speed must be below seven knots for conventional tugs to be effective, and it is a one-way position. A tug pushing on the port quarter will either oppose a turn to starboard or assist a turn to port.

This position does, however, have the advantage of leverage. Since it is well aft, the tug has a long assist lever. If the tug is on the appropriate side, this is an effective position to assist or oppose the ship's steering.

At slower speeds the tug can back on a line to retard the ship but, as with the forward tug alongside, the applied force will also apply a turning component to the ship. Again, two tugs backing on opposite sides of the ship would be required to balance their respective turning forces.

TUG ON SHIP'S TRANSOM.

A tug pushing on the ship's transom (fig. 19-4D) has the advantage of being able to apply steering forces to both port and starboard on the end of a long assist lever. It functions as a rudder, and tugs in this position are commonly referred to as rudder tugs. This is an excellent position to execute the assist or oppose response.

As in tugs alongside, a rudder tug pushing on the ship's transom creates a propelling as well as steering force. In addition, weather and transom shapes may make it prohibitive for a tug to push in this location.

RUDDER TUG TETHERED TO SHIP'S STERN.

This position (fig. 19-4E) is the domain of the tractor and ASD tug and is an optimal position to respond to a ship's steering or propulsion failure. The tug can direct steering forces either to port or starboard, readily change directions, and retard the ship as well.

The tractor or ASD tug has three options for applying the desired towline force and direction—direct towing, in-direct, and in-line arrest (fig. 19-5).

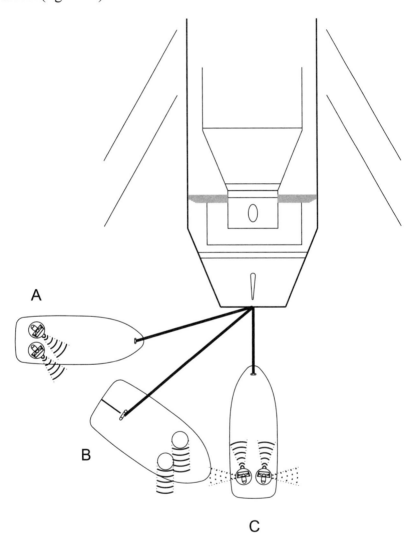

Fig. 19-5 Maneuvering options for a tethered omni-directional escort tug. A: Direct towing. B: Indirect towing. C: In-line arrest.

Direct towing is effective at slower speeds but rapidly loses effectiveness as ship speeds approach six knots (fig. 19-6A). Above six knots in-direct towing surpasses the efficiency of direct towing and generates higher steering forces as ship speed increases. Many of today's special purpose escort tugs are capable of in-direct towing of ten to twelve knots.

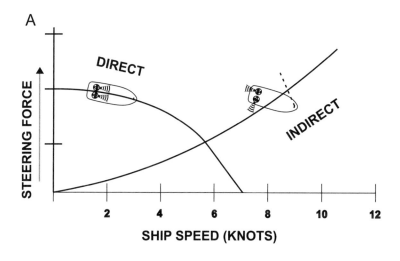

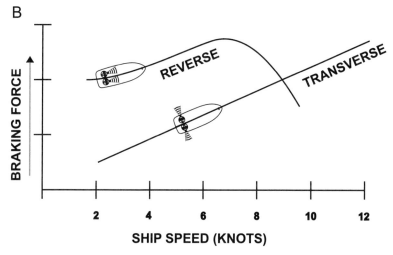

Fig, 19-6. Comparative effectiveness of omni-directional rudder tug maneuvers. A: Direct vs. indirect. B: Transverse vs. reverse arrest.

The in-line arrest position allows the tug to apply braking force without creating a turning component. The ASD tug has two methods of applying towline force in this position, transverse and reverse arrest. At ship speeds above seven knots the braking force of the transverse arrest option continues to escalate while the reverse arrest drops off rapidly (fig. 19-6B).

The rudder tug tethered to the ship's stern is in an optimal active escort and response position. It already is in a position to respond immediately to a ship malfunction. It can maintain the position at high escort speeds and can respond to a malfunction with a combination of braking and steering forces. It does not need to come in contact with the ship's hull.

SHIP SPEED—A CRITICAL FACTOR

The ship pilot's choice of escort speed has a direct bearing on the outcome of a ship's steering or propulsion failure. As one veteran pilot stated "When a ship has a steering failure, an extra knot or two of ship speed can make the difference between 'I know I can bring her back,' to 'I might get her back,' to 'there's no way she's coming back.'"

Ship speed can have a critical cause and effect on the three emergency response elements below:

Ship Advance and Off-track Carry. Ship speed predetermines the ship's advance and off-track carry should it lose propulsion or steering.

Compound Effects of Time Delays. Excessive speed compounds the effect of time delays in detecting or responding to the malfunction. At higher speeds the ship gathers more turning momentum in the same time increment and will thus take more corrective force.

Tug Response Positions. As noted, the tug response positions that require contact with the ship's hull cannot be attained or are effective at ship's speeds of greater than seven knots. In addition, ship speed has a direct bearing on the line and hull strains placed on the tug in indirect towing. The forces created by indirect towing are largely a result of the tug's hydrodynamic resistance. Hydrodynamic resistance increases as the square of ship speed. A difference of one knot in ship speed can exponentially magnify the line and hull forces placed on a tug responding by indirect towing. This magnification can mean the difference between the tug retaining its full capability or having to pull back to avoid deck immersion or equipment failure.

In summary, judicious management of ship speed is a prerequisite to sound ship escort practice.

TIME DELAYS

Time delays can undermine the best ship escort plan. The study and analysis of the intended escort route, the design and construction of state of the art tugs, and the extensive training in their use can all be for naught if there is too much delay in detecting and responding to a ship malfunction. Time delays can be a result of:

- Failure to immediately recognize the problem aboard the tanker
- Time taken to maneuver tug from passive to active escort position
- Time taken to connect the towline
- Time taken to maneuver the tug to the required response position and apply bollard pull
- Time delays due to tug equipment failure or malfunction

OPERATIONAL CONSIDERATIONS

Much of ship escort operations take place at speeds above that of normal ship assist work. These higher speeds have little tolerance for human error or mechanical failures.

Passing and Casting Off Towlines

Escort tugs may be called on to safely pass and secure a towline to the ship's stern at high speeds. If this task is not carried out in an efficient and competent manner it can be the source of a critical time delay or a fouled propulsion unit on the tug.

Generally, the tug is maneuvered in close to the ship's stern to pass or pick up the heaving and messenger lines used to connect the towline. The tractor tug is more difficult to handle than an ASD tug in these circumstances since the skeg on the leading end of the tractor is in the turbulent waters of the ship's propeller wash. Once the towline is passed, the tug pays out towline as it eases back into position.

Sea and swell conditions can make this maneuver difficult or impossible. Severe weather or escort routes that begin in open ocean waters may require a line throwing gun to pass the towline. The tug and ship's deck crew must exercise good and vigilant seamanship to prevent too much slack or the towline itself from slipping into the water and fouling the tug's propulsion drives.

Casting off an escort tug also is commonly done at high speeds. The ship's crew should wait until the tug has run up close to the ship, almost touching the ship's stern. The ship crew should slowly lower the towline to the tug, and only when the tug operator signals he is ready to receive the towline. This ensures that the towline does not fall into the water.

Tug Mechanical Failures

Ships being escorted can rely on an external resource, the tug, should they experience a mechanical failure. Tugs are not immune from mechanical maladies, but they must be prepared to call on their own internal resources to remedy the effect of mechanical malfunctions.

Both tractor and ASD tugs, so nimble with two functioning units, can be difficult to tame when one unit becomes inoperable. Tractor and ASD tug maneuverability arises from the ability to manage the balance of thrust between two propulsion units. When one unit fails that balance is lost. Although a well-designed ship escort tug should be able to provide some steering assistance if one unit fails, it requires a tug operator of the highest skill.

Human Factors

The technological advances in tug, winch, and towline design have added a new dimension to safety in shipwork and ship escort operations. As noted in chapter 6, the towline and its associated connecting gear traditionally were viewed as a weak link in towing. Although it may have resulted in a damaged ship or dock, a failed towline in many circumstances provided the safety net for the tug and its crew. The line would break before the tug ended up in peril. The synthetic lines and technologically marvelous but complex winches in today's state-of-the-art escort tugs have reversed this safety factor. In some circumstances it is more probable that the operator will fail before the towing equipment does. Some tugs in escort service have been pulled sideways, decks submerged to the point of capsize, yet the line and winch did not fail and release the tug. In these cases the equipment was functioning as it was designed, but the human behind the equipment either overdrove the tug or operated the winch improperly.

Both tug operators and ship pilots should be mindful of the fatigue factor in ship escort. The ship escort route often is of long distance and duration and may require extended periods of intense concentration. Running tethered to a tanker at escort speeds can be stressful, with little room for a lapse in concentration.

The consequences of an error in judgment or equipment failure can be immediate and grave at high escort speeds. Tankers are slow to change course. Should the tug running alongside the tanker experience a rudder or steering failure, the distance that seemed so ample immediately becomes inadequate.

The old adage of "life before property" holds true in ship escort towing as well. Both ship pilot and tug operator can place their escort plan and operations in that context by recognizing and managing these operational considerations.

STUDY QUESTIONS

1. What are the two primary purposes of tugs in ship escort?
2. What are the three tug response modes?

4. What is the least effective tug response position?
5. What is the most effective tug response position?
6. What are the three maneuvering options for a tethered ASD or tractor tug?
7. Why is ship speed critical to a ship escort response?
8. What is the danger of the new line and winch technology used in ship escort?

CHAPTER TWENTY

COMMUNICATION BETWEEN SHIP AND TUG

A critical element of successful shipwork is clear, professional communication. Good communication is the accurate exchange of information. This exchange has no universal protocol or vocabulary, but there are universal principles that attempt to ensure an accurate interchange of information.

Pilot and tug operator interchange has a twofold purpose in shipwork—one is to acquire information, the other is to initiate and complete tug actions.

Although the pilot ultimately maneuvers the ship and directs the tugs, he must acquire accurate information on which he can base shiphandling actions. Much of this information can come from the tug. The tug can relay the position of navigation aids, vessel traffic, and other navigation obstructions that the pilot may be unable to see. The tug operator may be able to make a more accurate assessment of the strength of a local current, the swing or set of the ship, or its proximity to a dock or bridge abutment. The perspective from the tug can be valuable in quickly conveying key information when an unforeseen problem develops (e.g., the ship crew just cast off the tug's line prematurely). Conveyance of such types of information can be essential. The pilot does himself and his charge a service by requesting the tug operator's input. The tug operator does well by offering information but respecting the pilot's authority and responsibility.

Communication that directs the tug's actions follows a "command and respond" dialogue. Today this dialogue is verbal, conveyed over VHF or other short-distance radios. In the past this conversation was carried out with whistles. The ship usually used its whistle to direct the after tug, while the pilot used a policeman's whistle to signal the forward tug. The tugs responded with their own whistles or sometimes one tug would use a smaller "peeper." This was done so the pilot or docking master could distinguish one tug from the other. This was an effort to avoid confusion about who was doing what.

The basic signals consisted of:

- One short blast means "come ahead on your engines" if stopped.
- One short blast means stop if you are *going ahead or astern* on your engines.
- Two short blasts mean back your engine.
- Three very short blasts is a signal for more speed.
- One prolonged blast is a signal for slow speed.
- One long blast followed by two short blasts is a signal to let go or shift the tug.

In some ports, notably New York, a fairly lengthy dialogue could be carried out between vessels using whistle signals. The whistle vocabulary in New York was more extended and complicated than the signals in general use.

Whistle signals had all the qualities essential to good verbal communication. They were simple, clear, easily understood, and not easily misunderstood. As tugs became more capable of complex ship-assist maneuvers the limited vocabulary of whistles had to expand to words. While whistles may have been appropriate for the conventional single-screw tug pushing or backing, today's dialogue of command and response requires a greater number and variety of descriptive terms communicated verbally over the radio. Instructions to the tug on the VHF radio should be precise. The tug should be identified first at each command and the orders given be as brief and direct as possible. Ambiguous terms or words with multiple meanings should be avoided.

The command and response dialect used between ship and tug varies by region, port, and the type of tug being used. Regardless of regional or local vernacular, the language must include clear terminology that:

- Identifies the tugs
- Indicates the direction of force to be applied
- Indicates the amount of force to be applied
- Indicates the desired push or pull angle
- Verifies the receipt, understanding and implementation of the command by the tug

Tug Identification

Various methods are used to refer to a particular tug when multiple tugs are used. Some refer to the tug by name, some by number to the tug's position relative to the ship (e.g., "number two tug, all stop"), or by a reference point on the ship (e.g., "bow tug ahead half"). All three are sound methods but pilot and tugs must have a mutual understanding and agreement on which methodology will be used. Many tug companies number their tugs (e.g., Cates I, Cates II, etc.), but this can create confusion if the pilot is referring to the tugs by numbered position rather than numbered name.

Direction of Tug Force to be Applied

Communicating direction requires a reference point and terms that describe motion towards or away from the reference point. Common reference points in shipwork are compass points, geographic points, and names of structures that are close by and easily understood (e.g., "easy towards the dock"). As the ship enters more constricted waters or approaches the dock, a pilot may change the reference point used to communicate direction. He should make these transitions clear to the tug. In addition, pilots should avoid the use of words that sound similar. "Toward" and "away" from the dock are more distinct than "on" and "off" the dock.

Amount of Force to be Applied

Typical power commands are increments of the tug's total available horsepower, towline loads (if the tug is equipped with a tensionometer), or specific engine rpm. Unless the pilot is familiar with each tug's specific engine and associated propulsion system, the use of engine rpms should be avoided. 800 rpm on a slow-speed diesel engine has quite a different result than 800 rpm on a high-speed engine.

Most pilots use a power terminology that divides the tug's total available horsepower into increments (e.g., slow, one-third, two-thirds, etc.) Again, as in directive commands, terminology should be chosen that is not confusing to the ear. One of the most easily confused terms is "full," which sounds similar to pull. "Pull towards the dock" can easily be mistaken for "full towards the dock."

Desired Angle of Push or Pull

In many situations the pilot desires to adjust the angle of the pushing or pulling force applied to the ship. Changes in pushing angles can be communicated with terminology that refers to the tug's aspect to the ship's hull or other easily understood reference point (e.g., "angle more to the north").

Directions to change a towline angle can be more ambiguous. The shiphandler familiar with tugs knows that it is the towline angle coming off the ship that indicates the direction of force, not necessarily the angled aspect of the tug. A tractor tug in the indirect towing mode is a clear example of this. In these cases it is best for the pilot to communicate the desired angle of the towline force as opposed to the angled aspect of the tug.

Verification of Tug Response

A key axiom of good communication is that the response is as important as the command. The tug operator's response must clearly indicate that he has understood and carried out the pilot's directive. The response should be timely, comprehensive, and precise. It must include both the tug's name and a repeat of the pilot's order (e.g., "Andrew, away from the dock, one-half."). Responses such as "roger" or "affirmative" do not give the pilot the necessary confirmation that the right tug understood the right order.

Communication with Tractor and ASD tugs

The introduction of omni-directional tugs has added new dimensions to pilot-tug communication. These tugs have multiple methods at their disposal to apply force to a towline and the ship. In addition, the amount of applied bollard pull reflects a balance between engine rpm, angled alignments of the propulsion units, and hydrodynamic lift or drag of the tug. These variables can be manipulated by the tug operator in multiple ways to provide the same effect. As an example, the directive "push toward the dock slow" given to an ASD tug can be accomplished in multiple ways. The tug operator can have one drive unit clutched in, aligned parallel to the tug's centerline, pushing directly towards the ship, or have both units engaged, but "feathered" in at about a 45°. Both alternatives have the same result.

Because of these maneuvering variables and the complexity of omni-directional tugs, the interaction between pilot and tug has become more of a contributing partner, rather than a command-and-respond type of communication. The interest of the pilot lies in the desired result from the tug's force. The type of towing mode (direct, in-direct, powered in direct, etc.) is made at the pilot's discretion but in collaboration with the tug operator. The tug operator has the better expertise to choose the appropriate engine and azimuth configuration to produce the pilot's intent.

Different pilot groups and the tug operators with whom they work have different approaches to finding the balance between communicating what needs to happen and how to make it happen. Some groups have implemented a standard set of terms, while others rely on a more fluid interaction in the terms used. Similar to the past tradition of whistle signals, today's directives to omni-directional tugs follow the same path of some shared, some unique terminology by region and port.

COMMUNICATION DYNAMICS

Interchange between pilot and tug operator is more than just verbal. Actions as well as words communicate information.

The tug operator communicates the tug's position and force by the consistency and precision of his response to the pilot's orders. The pilot uses the ship's motion to gauge the effect of the tug and gain a feel for the ship. If a tug operator responds to a command of dead slow with half the tug's rated horsepower, his actions are communicating false information. The pilot earmarks the effect of the tug at dead slow and may erroneously base future commands on the assumption he still has most of the tug's horsepower in reserve. Similarly if a tug continues to push on the side of the ship to maintain position, even after the command "all stop" has been acknowledged, the pilot may mistake the tug's unreported push for an odd current or eddy.

Anticipatory action is another means of non-verbal communication. This is a characteristic of tug operators and pilots who have worked together for a long time. For example, when the tug is pushing toward the dock and receives the command to stop, the operator stops, stretches, but does not weight his line in anticipation of the next order, "Back easy away from the dock." Likewise, the pilot can break down the commands for the same sequence of events. "Stop", "Stretch your line," "Back easy away from the dock" communicate to the tug operator that the pilot is maneuvering within the limits of tug and ship. Each has such a good mutual understanding of the particular shipwork maneuver in which they are engaged that verbal commands and responses merely confirm what both parties already know.

Nevertheless, professional mariners do not make the mistake of discarding communication procedures even if they seem superfluous. They know that circumstances can change at any time, and that even a maneuver they have done one hundred times without incident can one day depend on clear, concise, and precise communication.

STUDY QUESTIONS

1. What are the three qualities of good verbal communication?
2. Why did the verbal radio communication replace whistle signals?
3. What terminology must be clear for effective command and response communication?
4. Why should the terms "full," "on," and "off" be avoided?
5. How does the tug acknowledge an order?
6. What is a means of non-verbal communication in shipwork?

CHAPTER TWENTY-ONE

CONCLUSION

The call for tugs to assist ships continues to grow worldwide. New tug construction has been flourishing as a result of the demand for improved tug performance and the sheer number of tugs required by this expansion. In many ports, particularly in the United States, one generation of tugs is being created while another is retired. The majority of the new-generation tugs are tractor and ASD tugs dedicated to shipwork.

As tug fleets have grown and new tonnage replaced old, so too have the requirements for skilled tug operators and shiphandlers. Today's tugs are more complex, sophisticated, and capable than their predecessors, yet the principles of their employment and effectiveness remain as they have been since they began towing vessels in 1802.

The shiphandler of centuries ago would be impressed with the advances in size, power, and technology in tugs and ships. Yet, if he were transported to the bridge of a ship today, his shiphandler's eye would recognize the same shiphandling principles at work—the application of the intrinsic power of the ship and the external power of the tugs, as effective levers to maneuver the ship safely to and from its berth.

He also would recognize that the successful application of these principles continues to rest squarely on the shoulders of the operators of the two vessels. Technology has given both the shiphandler and the tug operator exceptional tools but it has not diminished the dependence on the human element.

The most able tug still requires the knowledge and experience of two professional mariners to use its full potential in shipwork. Both tug operator and shiphandler must be vessel-handling artists in their own right.

In reality the interaction between the tug and the ship is a collective effort, like that of an orchestra. The pilot or shiphandler is the conductor. The success or failure of the performance is contingent upon several factors: the knowledge of the shiphandler, the quality of the instruments (both ship and tug), and the skills of all of the performers, including the tug's master and crew and the officers and crew of the ship. All participants must have a mutual respect for, and understanding of each other's authority and expertise.

As in any art, practice is essential in keeping shiphandling skills and senses sharp. Whether tug operator or shiphandler, those that have the vessel handling touch, maintain it through practice and a continual learning process. These professionals are dedicated to transforming their intellectual comprehension of the new tools and techniques of shipwork into a hands-on understanding.

Shipwork is a timeless business. Although tug and ship technology continually advance, the cornerstone of success in shipwork remains with the professional mariners whose calling is the art of shiphandling with tugs.

ANSWERS TO THE STUDY QUESTIONS

CHAPTER 3—THE TUGS

1. Conventional, Tractor and Azimuthing Stern Drive Tugs.
2. Towing Point, Propulsion Point, Propulsion & Steering, Maneuvering Lever, Hull Shape, and Superstructure and Fendering
3. The last physical point on the tug that fairleads its line; or the contact point between tug and ship when the tug is pushing.
4. The focal point of the tug's application of horsepower under the water.
5. The distance between the towing point and the propulsion point.
6. Conventional Tugs- towing point forward of propulsion point, maneuvering lever short
   Tractor Tugs-Towing point aft of propulsion point, maneuvering lever moderate
   ASD Tugs- Towing point forward of propulsion point, longest when towing point is the bullnose- shorter when the towing point is the tow bitt.
7. When the tug does not have enough leverage to counter the pull of the towline.
8. Efficiency, since on a horsepower for horsepower basis, a single-screw tug will develop about 20 percent more thrust.
9. Lack of maneuverability, inability to steer astern, and inability to maintain position (i.e., 90°) when backing. Power astern is less than power ahead.
10. Excellent maneuverability, ability to maintain position when backing (if ship is stopped), and ability to steer astern.
11. May require a stern line when backing, is less efficient than a

single-screw tug, and backing power is less than power ahead.

12. A tractor tug can steer when going astern, has omni-directional thrust and does not have the same risk of capsizing as a conventional tug when towing.

13. Better sea-keeping ability, shallower draft, more efficient in converting horsepower to bollard pull and can perform a transverse arrest.

CHAPTER 4—PROPULSION AND STEERING

1. Power Source, Power Transfer & Control System, Hydrodynamic Driver, & Thrust Directional Control Mechanism.

2. To transfer and control rotative force to the tug's hydrodynamic driver (typically a propeller).

3. Allowed smaller, more powerful tugs to be built, increased the range and power of tugs.

4. The DR system is the simplest system of maneuvering. The engine is directly connected to the propeller shaft, and the direction of its rotation is the same as that of the engine. The engine must be stopped between maneuvers or when changing maneuvers from ahead to astern and vice versa. The engines are started by compressed air injected into a reservoir cylinder. These engines may be fitted with wheelhouse controls so that they need not be tended by an engineer.

5. Limited number of maneuvers, engine speed may not permit a propeller of the most efficient size, vulnerability to mishap (i.e., failure to start, or starting up in wrong direction).

6. The DE system consists of engine-driven generators that supply current to the electrical-drive motors connected to the propeller shaft. The direction and speed of the drive motors is determined by the control of the current developed by the generators.

7. A wide range of speeds available from stop to full speed ahead or astern. The fact that the main engine (or engines) runs at a constant speed.

8. High costs, vulnerability to salt, and dampness.

9. The SCR system employs an AC generator whose current is then converted by silicon control rectifiers (SCRs) to DC current for driving the propulsion motors.

10. The SCR system is a more modern development of the DE drive. It is less costly than DC installations, and the main engine generators can furnish current for ship supply.

11. The controllable pitch propeller is not only an energy converting (rotative force to thrust), but it is also a transmission since it changes the direction of thrust by reversing the pitch of the propeller blades. The amount of thrust is also regulated by reducing or increasing the pitch as required by the circumstances. Stopping is achieved by flattening the pitch so that it is in neutral and does not develop thrust ahead or astern even though the propeller continues to turn. The pitch is controlled by an inner shaft inside the drive shaft that engages cams on the propeller blades in the hub of the drive shaft.

12. Amount of thrust and reversing mechanism.

13. High costs, susceptibility to damage. A vessel's steering qualities are affected when in neutral pitch. Since the propeller is always turning, it is liable to foul lines in shipwork.

14. Through sets of additional gears and clutches; one for reverse and one for ahead.

15. Quality and velocity of water flow; and number, pitch, shape, size, and speed of the rotating propeller blades.

16. The propeller turns within a shroud which increases thrust 15 percent to 60 percent more than an open propeller.

17. They present more drag and lateral resistance underwater, hindering maneuverability and light tug running speed.

18. Velocity of water flow, quality of water flow and rudder shape.

19. Balanced or semi-balanced rudders are constructed so that the leading edge of the rudder extends forward of the rudder post. This is done to provide more effective steering and a mechanical advantage to the steering gear.

20. A "spade" rudder is one that is not supported at the bottom by an extension of the vessel's keel or skegs.

21. A steerable nozzle changes the angle of thrust by rotating the angle of the nozzle around a fixed propeller. A steerable propeller changes angle of thrust by rotating the propeller around a vertical axis.

22. Flanking rudders are installed ahead of the propellers, and there are usually two of them for each propeller. They are more effective when operating astern as they are on the discharge side

of the propeller.

23. The VSP functions like a controllable pitch propeller rotating around a vertical axis. Vertical blades are attached to a rotating rotor casing. The rotor turns while each vertical blade oscillates, changing its angle (pitch) at points along the circumference of the rotor casings track.

24. VSP horsepower to bollard pull ratio is lower than conventional or steerable propeller tugs; VSP tug will have a deeper draft than a conventional or ASD tug, VSP systems are expensive compared to a steerable propeller or conventionally propelled tug.

25. Engine power is transmitted through a shaft to the upper casing of the drive unit housed inside the tug. A pneumatic or hydraulic clutch regulates the connection between engine output shaft and drive unit input shaft. A series of right-angled gears and shafts converts the horizontal rotation of the engine output shaft into the horizontal rotation of the propeller shaft. The lower drive unit houses the propeller and can be rotated through 360 degrees by mechanically or electrically driven hydraulic motors

26. Reversing thrust requires rotating the drive unit 180 degrees; the SPS may be constantly thrusting while shifting thrust direction.

27. Bollard pull is the pull in pounds or kilos generated by a tug pulling against a fixed object. In conventional tugs it is estimated to be about one long ton (2,240 pounds) per 100 hp, depending upon the configuration of the tug and whether or not nozzles are installed. Bollard pull can range between 22.5 and 38 pounds per hp.

28. Not always since other factors, like "stiffness" and maneuverability, are important too.

CHAPTER 5—STRUCTURAL CONSIDERATIONS

1. A tug for shipwork will have a heavily constructed hull with low freeboard. It should be "stiff enough" so that a strain on a line leading to the side will not capsize it. The bulwarks are "tumbled home" so that they will not be damaged when the tug goes alongside a ship even if they are both rolling a bit.

2. Yes, the bow of a conventional or ASD tug is heavily reinforced since this is the area that comes in contact with the ship when the tug is pushing full ahead against the vessel. The bow

in powerful tugs should be of ample radius so that large fenders can be used to distribute the tug's force over a large area of the ship's side. Heavy rub rails are usually installed along the tug's sides, fore and aft, above the waterline to reinforce and protect the hull. In tractor tugs the stern is usually strengthened since tractor tugs normally push stern first against the ship.

3. The deckhouse and/or wheelhouse on a harbor tug is set inboard to prevent it from coming into contact with the ship's side when the tug is working under the flare of the ship's bow or overhang at the stern and when the two vessels are rolling.

4. A "bullnose" is a large closed chock installed near the bow of a tug to provide a fairlead for the tug's head line and springline. It should be large enough for the large working lines and the splices to pass through freely.

5. The staple functions as the tug's towing point and is subject to extraordinary strain. The staple is the primary focal point of a tug's bollard pull when on the towline. The staple itself must be heavily constructed and mounted in a manner that solidly connects it to the tug's internal frames, bulkheads, and plating.

6. There may be a large H-bitt installed fore and aft in the foredeck, another H-bitt installed athwartships aft of the deckhouse, and quarter bitts installed or augment the forward and aft H-bitts.

7. The bitts and deck fittings must be large enough on which to belay the large size working lines, and strong enough to withstand the heavy strains imposed upon them. The "bullnose," forward cruciform bitt, and the quarter bitts must be set inboard enough so that they will not come in contact with the ship's hull and cause damage.

8. A "dipping" mast should be installed so that it can be lowered when working under overhanging obstructions on the ship.

9. The wheelhouse should be constructed in such a fashion that the tug captain's vision is as nearly unobstructed as possible. The controls for the helm and engines should also be located so that his view is unobstructed.

10. Because they are less likely to do damage if they come into contact with the submarine's hull than either twin-screw or chine-(V-bottom) hulled tugs.

11. Poorly located chocks and bitts installed too far forward or aft or beneath overhanging structures can damage or endanger the tug. External hull fittings on the ship's side (i.e., sponsons, pipes, and rub rails) can also be a nuisance or a hazard.
12. "Pocket chocks" are set into convenient locations in the ship's side in such a way that they have a bar or pin for the tug's lines to be secured to. They are particularly helpful in the case of some passenger vessels, container ships, and car carriers that have short forward and afterdecks.
13. Absorb and dissipate energy; enhance or reduce traction; create spacing between tug and ship.

CHAPTER 6—GEAR AND RIGGING FOR SHIPWORK

1. Working lines, Fastening/Release Mechanisms, Best Practices
2. Nylon, Polyethelyne, PolyPropolene, blends, HMPE and Aramid fibers.
3. Not as a stand alone line—they are much too elastic. However, short sections of nylon line may be used in composite lines.
4. Slings, mooring and working lines.
5. Light work where a floating line is best.
6. In all applications because they are stronger than polypropylene and polyethylene, but are lighter than Dacron, and they also float.
7. Spectra® and Dyneema®
8. They protect the working lines from chafe where they pass through the chocks on the ship. They are usually between 15 and 25 feet in length.
9. 3, 6, 8, 12 strands and Double Braid.
10. 6x19 or 6x37 construction made from improved plow steel or extra improved plow steel.
11. Abrasion resistance, disposable component, accommodation of dynamic loading, creation of a weak link, provide flexibility and softness to working ends.
12. Braking capacity, Maximum line pull capacity, Slack line speed, Remote release
13. Choose the right rope for the right job. Use the rope properly
14. Heaving lines, tag lines, goblines, and quick-release straps.
15. They are usually about 15 fathoms in length and inch to ½ inch in diameter to provide a handhold for the deck force on the

assisted vessel. A "monkey's fist" or other form of weight is secured to the outboard end to facilitate throwing.

16. A tag line is a length of line 30 to 40 feet in length about 1 inch in diameter to provide a handhold for the deck force of the ship. It is attached to the outboard end of the working line or wire rope pendant and is intended to facilitate heaving it aboard.

17. Yes, manila is suitable for this purpose since it provides a good grip. Slippery lines like polypropylene should be avoided.

18. A gobline is a stout line used to tie down the towline of a tug when the tug is doing a towline job. Its purpose is to secure the towline nearer to the stern of the tug and restrict its arc of movement so that the tug will not be girt or tripped. It is especially useful when the tug is being towed stern first.

19. A quick-release strap is a relatively short length of line secured to the after H-bitts on a tug and that is rove through the eye of a hawser used as a towline. Its purpose is to provide a safe method of "slipping" the towline without endangering the crew of the tug.

CHAPTER 7—HANDLING THE LIGHT TUG

1. A point in a maneuvering sequence that requires a critical action at a specific time.

2. A "right-hand propeller" will have a tendency to move the stern to starboard and the bow to port when engaged ahead. When reversed, it will move the stern to port and the bow will fall off to starboard.

3. It is steered when moving astern by giving an occasional "kick" ahead on the engine to correct the heading. This is done briefly enough so that it does not destroy sternway.

4. It is turned by "backing and filling," i.e., alternately going ahead and astern on the engine with the helm turned toward the direction of the turn.

5. Usually it will since the stern will be cast to one side or the other when the tug backs its engine. This will be to port with the standard right-hand propeller. Tugs with nozzles may not react the same way.

6. They enable a single-screw tug to steer astern and maintain position when backing during shipwork.

7. No, because when the tug is backing and filling to make a tight turn, the nozzle angle must be reversed between the ahead and astern engine maneuvers.

8. Both propellers thrusting in the same direction with the same rpm will balance each other's torque and off-center position. By manipulating the balance between the two engines, a twin-screw tug can steer with its engines.

9. That the tug is being maneuvered with one engine ahead and the other astern. The tug can be turned completely about without moving ahead or astern.

10. Flanking means moving a tug laterally by using its engines and rudders in opposition. For example, if the tug's port engine is maneuvered ahead with the rudder to port and the starboard engine is backed, the tug will move sideways to starboard.

11. By coming ahead on a springline until the stern is open and then backing clear of the dock. However, a twin-screw tug may "flank" off of the dock.

12. It will usually approach the dock at an angle of 15 degrees to 20 degrees, stop its engine, then bear off, and back its engine when close alongside. It can then be worked ahead on a spring-line until it is in position.

13. The wheel controls the bow and the pitch levers control the stern. The pitch levers also control fore and aft movement of the tug.

14. into

15. One is by using pitch levers alone (reversing pitch); the other is a combination of pitch levers and wheel (turning the tug 90 degrees to its initial heading).

16. Flanking a VSP tractor tug is done by turning the wheel in the desired direction of lateral movement; setting the outside lever (the one closest to the direction of desired movement) forward and the inside lever aft. Fine adjustments between the three controls will propel the tug directly sideways.

17. When holding the tug's bow off the dock the wheel-induced transverse thrust may prematurely retard the tug's way.

18. No, they always point in the resultant direction of the azimuthing drives thrust. When moving bow first, the drive units are rotated in the direction opposite of the desired turn. This is

because the stern is steering the bow. When steering while moving stern first the drive units are rotated in the same direction as that of the desired turn.

19. The outboard drive unit (one closest towards the direction of lateral movement) is in a "backing" alignment while the inboard drive unit is angled ahead and in. The azimuth and rpm of each drive unit are adjusted to produce the desired effect.

20. The tug operator must anticipate the path of the tug and the drive unit configurations required during the sequence of the maneuver.

CHAPTER 8—SHIP–TUG INTERACTION AND TUG HANDLING

1. Tug hull resistance, opposing forces, application of force to ship
2. Ship Speed, displacement, and hull shape
3. High pressure at the bow, lower pressure alongside the ship and suction at the stern.
4. It draws the tug toward the ship by the suction aft and it pushes the tug away from the ship forward.
5. By pacing the ship long enough to see how the tug is affected and then gently easing the tug in alongside. An experienced operator lets the tug be drawn in slowly by the suction aft with the helm turned slightly away from the ship, works the tug in against the hydraulic forces forward
6. Stemming" can occur when a tug takes a towline from the bow of a moving ship. If the tug comes in contact with the ship's side it may lose steerage, in which case the tug may be caught by the ship's bow and rolled over.
7. Misjudgment of the tug's set in towards the ship's bow or oversteering as the tug transits through the ship's pressure zones at the its bow.
8. The tug operator should avoid backing when the tug is overtaking the ship to get into position to push or pull. When the tug backs its engine, the tug may not respond and can cause damage to itself and to the ship.

CHAPTER 9—BASIC TUG POSITIONS IN SHIPWORK

1. Lift and Drag
2. The ship's mates on the bow and stern should direct their deck crews to slowly lower the tug's line and tagline unless directed

to drop it by the tug's operator. Dumped lines have potential to cause serious injury to the tug's deck crew as well as foul the ship or tug's propulsion units.

3. The tug's towing point is forward of the propulsion point which can make the tug susceptible to tripping or girting.

4. Tripping or girting the tug can be caused by the ship overtaking the tug, or when the tug's thrust and maneuvering lever cannot counteract the force of the tug's center of hydrodynamic pressure acting on the tugs turning lever.

5. By employing tractor or reverse tractor tugs or requiring that the ship go at slow-to-moderate speeds.

6. It takes less time for the ship to respond to the tug's efforts, and working the tug alongside is safer than towline work.

7. The conventional tug uses a quarter line to maintain a position approximately at right angle to the ship when backing, otherwise the effect of the torque of the tug's own propeller, the vessel's motion, or wind and current might move the tug out of position.

8. A wrap line is line used by a tug to steer the bow of a ship moving stern first. It is led from one side of the ship's bow to the bullnose on the tug located on the opposite side of the ship's bow. This allows the tug to steer the bow in either direction.

9. No. Bulbous bows, severely raked stems, and ships with flare may prevent a tug from working in this position.

10. This occurs when a tug with a ship or barge in tow astern makes a hard turn and at the same time applies more power. Since the tug's rudder will set the stern of the tug in the opposite direction, the bow of the towed vessel will often also be set in the opposite direction of the turn at first.

11. If there is too much slack in the lines, the tug's angle to the ship's centerline will vary and detract from the tug's ability to apply effective steering forces.

12. Indirect towing uses the tug's hydrodynamic resistance to add force to the towline.

13. Indirect towing can be used to provide steering and braking forces at ship speeds that are too high for effective direct towing.

14. Because the momentum of the tug moving astern could develop enough force to part the line.

15. The propeller may cavitate.
16. If the tug is overtaking the ship and backs its engine, the tug may not respond and can cause damage to itself and to the ship.
17. Human error, mechanical failure, and weather.
18. Human error in regards to excess speed, because if it is on the part of the ship, it can overpower a tug on a towline and trip or girt it. Excess speed can also capsize a tug using a quarter line and even endanger a tug fast alongside with only a head line out. It will also destroy the tug's ability to assist the ship. When the tug uses excess speed, it can part lines (injuring personnel) and lead to collisions with the ship it is assisting.

CHAPTER 10—BASIC SHIPHANDLING PRINCIPLES

1. Center of Lateral Resistance, Pivot Point and Maneuvering Levers.
2. Lateral vs. rotational
3. • Stopped—amidships at the CLR depending on draft and trim.
   • Moving laterally—amidships at the CLR depending on draft and trim.
   • Moving ahead—approximately one-quarter ship length aft of the bow
   • Moving astern—approximately one-quarter ship length for ward of the stern
   • Turning—approximately one-third ship length aft of the bow
4. Fulcrum

CHAPTER 11—STEERING AND PROPELLING SHIPS WITH TUGS

1. The distance between the location of the tug's applied force and the ship's pivot point.
2. The ship has greater power and can apply that power on a longer and more efficient maneuvering lever. This can easily overpower the tug.
3. Push, act as a drogue or pull on a line.
4. Angle of the tug, offset position of the tug, drag force and tug's applied horsepower.
5. At all but the lowest speeds the conventional tug allocates almost all its horsepower to maintain position at, or close to, 90°. There is minimal lateral pull applied to the tug's headline. The

tug's angle, offset position, drag, and horsepower are all working on a long assist lever to counter the turn.

6. Twin screw.

7. Steerable bow thruster.

8. Negative water flow is when the tug's wash opposes the general
direction of the water flowing by its hull.

9. Negative flow considerably increases the torque loadings
on the propeller and engine. High negative water flow may
increase the load enough to stall, or damage, the tug's engine,
clutch, or couplings.

10. A trailing tug creates the longest assist lever and allows the tug
to apply equal forces to both starboard and port. The tug can
enhance the effect of the ship's rudder on a ship with headway
and can also control the bow and steer a ship with sternway

11. The shiphandler must execute careful management of ship
speed and propulsion wash.

12. A tractor or ASD tug can utilize direct towing at ship speeds of
five knots or less, and indirect towing at higher speeds

13. Steering and propelling ships with tugs is a matter of creating
and using the tug force as an effective lever in conjunction with
the ship's maneuvering lever.

CHAPTER 12—TURNING AND LATERALLY MOVING SHIPS WITH TUGS

1. No, the pivot point may coincide with the CLR when the ship
is stopped, but once a lateral force is applied the pivot point
moves in response to the applied force.

2. The pivot point moves away from the tug end of the ship toward
the ship's stern.

3. Approximately two ship lengths

4. The pivot point will remain approximately midships.

5. Equal and opposing leverage

6. Drift angle is the angle between the ship's centerline and the
direction in which the ship's bridge is actually traveling in a turn.

7. Drift angle serves as a visual cue for the pilot to estimate the
degree of lateral motion, or slide, that a ship has as it executes
a turn.

CHAPTER 13—CHECKING A SHIP'S WAY WITH TUGS

1. Drag of the tug, direct bollard pull, or transverse arrest

CHAPTER 14—BASIC SHIPHANDLING MANEUVERS WITH TUGS

1. Initial assessment, create a maneuvering plan, adjust the plan
2. Sequence of maneuvering events, safety of tug, limits of tug design, skill level of tug operator, need and location of highest horsepower tug, required allocation of tug's horsepower to maintain or achieve assist position

CHAPTER 15—HANDLING DEAD SHIPS

1. With a "dead" ship the tug is the prime mover, with a powered vessel the tug is merely assisting.
2. Yes, just as tugs frequently handle fairly large barges unassisted.
3. The tug should be hipped up on the inboard side of the turn, since even if the tug backs the ship will continue to swing in the direction of the turn. If the tug (especially a single-screw tug) is on the outboard side of the turn, it may not be able to back as this would "kill" the swing.
4. If the ship's turn rate is too slow and there is neither sufficient room ahead for the ship to advance or to the outboard side for the ship's stern to rotate around its pivot point.
5. Yes, this is caused by the offset location of the propelling and steering forces.
6. Yes, the number of tugs required (unless stipulated by port regulations) will depend upon the size of the ship and the horsepower of the tugs.
7. Yes, but extra care should be taken when docking to avoid damaging the rudder and propeller
8. With two tugs: a powerful tug aft on the hip to propel and a smaller tug forward for steering placed on either bow. With three tugs: a powerful tug aft on the hip with two smaller tugs forward on each side for steering, or two tugs aft with one tug forward for steering. With four tugs: two tugs forward and two aft on the hip for very large ships.
9. Because two moderately powered tugs can propel a sizeable vessel since both tugs are moving the ship. The forward tug can steer the vessel, and the after tug can back to check its way. This is an efficient use of power especially for longer shifts.
10. By using the tugs as if they were components of a twin-screw ship and by using their engines in opposition for steering effect.

CHAPTER 16—TUGS AND ANCHORS

1. Because when a vessel uses its anchors, it can dispense with the services of a tug. There are also occasions when the use of the anchors is preferred.
2. The anchor is essentially a "passive" device that cannot push nor pull like a tug unless it has been "spotted" beforehand and is used as a "deadman" to heave against, or is used as a "brake" to check a vessel's way.
3. Restrain but not stop the ship, fix or hold the anchored end of the ship, or provide a deadman to heave against.
4. The anchor is used to provide directional stability when the ship is towed by the tug and to fix an end for the tug to turn the vessel.
5. Stern anchors may be used to keep ships from swinging with the tide, and are often used on passenger ships to hold up the stern of a vessel to provide a lee for the launches. When used with tugs they are often used in docking in fair tide situations.
6. The possibility of the ship overriding the anchor and as a result doing damage to the rudder or propeller.

CHAPTER 17—USING TUGS IN EXPOSED WATERS

1. The tug should be very heavily fendered and its working lines should be stronger than those required for normal harbor service.
2. Large pneumatic fenders (like tires) that are set on axles in a well in the bow of the tug. They are useful for shipwork in exposed areas since they will roll if the tug is pitching heavily as the tug pushes against the ship
3. Work the tugs on the lee side when possible and give the lines a longer lead to prevent parting them in the rise and fall of the swell
4. No, it takes the tug longer to respond to helm and engine in a seaway than it does inside a bay or harbor.
5. It usually puts up a line on the lee bow clear of the chocks where the mooring lines are passed and assists the vessel to hold its position while the gear is being run. The tug can also assist the steering while the vessel is making its final approach and backing to stop.

CHAPTER 18—SALVAGE AND EMERGENCY SITUATIONS

1. The resistance of ground effect (the weight of the vessel resting on the bottom) times the coefficient of friction of the bottom and the effect of wind, current, and heavy seas.

2. • Wait for a high tide to float vessel.
    • Discharge ballast or fuel.
    • Discharge cargo into lighters.
    • Pull vessel off with tugs.
    • Heave vessel off with beach gear.
    • Jettison some of the cargo.
3. Jettisoning the cargo is considered to be the last resort since the vessel's cargo is sacrificed to save the ship.
4. The vessel's draft should be checked to see how much buoyancy it has lost; soundings should be taken around the perimeter of the ship's side to determine what portion of the hull is aground; and the bilges should be sounded to determine if the vessel is holed.
5. Only if there are enough pumps or compressors to keep it afloat.
6. The buildup of sand or mud around a grounded vessel that will make refloating it difficult.
7. They can force a vessel farther aground and hamper salvage operations. Conversely, they can also help refloat a vessel by breaking the bottom suction or grinding down soft coral and limestone bottoms.
8. Lead out its anchors to restrain it, or ballast it down until the seas subside or assistance arrives.
9. The vessel may require assistance steering to avoid broaching when it is refloated.
10. A small tug can take a strain on the ship that will prevent it from broaching or "climbing the beach" until more assistance arrives.
11. "Beach gear" consists of heavy wire rope falls that are set up by a winch. They are attached to a heavy cable that is in turn attached to a shot or two of heavy chain and an anchor (usually an Eell anchor) which is spotted offshore of the stranded vessel. This gear was often set up on the ship's deck or on a barge attached to the ship, but modern salvage tugs sometimes are equipped to place the anchors and are attached to the ship by their own tow cable.
12. Yes, since they can apply their force from several different directions and "swing" the ship back and forth while beach gear can only pull in one static direction.
13. To assist in beaching a vessel that has been holed, remove a

burning tanker from an oil dock, breast a listing ship to the dock, and assist a vessel that has lost power or steering.

14. First, the tug should have its own crew in fire-fighting gear or oilskins and have hoses led out and ready to provide some protection to the tug's crew. Then, the tug should back down and connect its towing gear to the vessel's fire warps and take the vessel easily in tow after streaming about 250 feet of towline or cable.

15. Usually, but the vessel will have a tendency to sheer away from the tug. It will be more easily controlled if there is another tug on a towline or pushing at the other end of the ship.

16. By taking the vessel in tow astern or on the hip.

17. By handling the ship with a towline to the vessel's bow; making fast alongside of the vessel forward; and by steering the stern of the ship.

18. Stay well to weather of it, or clear out!

19. The safety of his crew, the safety of his vessel, and the safety of the vessel being assisted. His crew's safety should not be risked unless lives are at stake.

CHAPTER 19—SHIP ESCORT

1. Emergency Stopping
   Emergency Steering
2. Retard
   Assist
   Oppose
3. 2 tugs, aft of the ship's bridge
4. Forward shoulder of ship
5. Rudder tug tethered to ship's stern
6. Direct, indirect, in-line arrest
7. Determines advance and carry
   Compounds effect of time delay
   Determines effectiveness of response positions
8. The line may not break or release if the tug is in peril

CHAPTER 20—COMMUNICATION BETWEEN SHIP AND TUG

1. Simple, clear, easily understood not misunderstood
2. More complex tugs and maneuvers required a bigger vocabulary than whistle signals could provide.

3. Communication that:
   - Identifies the tugs
   - Indicates the direction of force to be applied
   - Indicates the amount of force to be applied
   - Indicates the desired push or pull angle
   - Verifies the receipt, understanding, and implementation of the command by the tug
4. They can be mistaken for other words
5. The tug operator's response includes the tug's name and repeating the order back to the pilot
6. Action of the tug

APPENDIX

The appendix that follows is drawn from technical papers and studies of tug performance, and are based on trials or ongoing practices. This material is included because it is relevant to the shiphandling process and is not readily available to professional mariners. The subject matter is selectively confined to those portions of the text that are likely to be of interest to the shiphandler.

Readers interested in referring to the complete texts for technical reasons should contact those who have generously made portions available for publication in this volume.

SHIP HANDLING AT RAS TANURA SEA ISLAND

Whereas this paper is officially entitled "Ship Handling at Ras Tanura Sea Island," during its preparation, I decided to broaden its scope in order to reflect our experience dealing with other large ports. The title should be changed to "Rationalized Design Parameters for Ship-Assisting Tug Boats" or "Study of Tugboats for Port Hypothetical".

If one were to attempt on a worldwide basis to cover the entire spectrum of techniques used by ship-assisting tugboats, describing in detail their methods, the diversity of procedures would produce a huge volume on the subject.

Virtually every seaport has its own features, peculiarities and problems. The methods used at each port have evolved through experience and generally in an empirical manner. Tugboat sizes and characteristics have kept pace with the increase in size of the vessels calling at that port.

This paper is an attempt to provide a broad background of information concerning ship-assisting tugboats and then to provide a rational approach toward recommending the type, size, and thrust requirements as well as the number of tugs required for Port Hypothetical. This method has been used and tested in determining the requirements for a major Alaskan oil port as well as a major Arabian Gulf oil port.

In this computer age, naval architects turn to the computer to generate the characteristics of their vessel designs in conjunction with their own personal training, experience and skill. This new design tool has furnished a refinement in the design process which was out of reach heretofore.

SHIP ASSISTING METHODS

Broadly, there are two distinct methods of assisting ships by the use of tugboats. First is the "towing on the hook" method as generally practiced in Europe and known as the "European" method. The other is the "towing alongside" method which is generally practiced in the United States and known as the "American" method.

The "European" method requires at least a head tug and a stern tug connected to the ship's bow and stern, each with a tow line attached to the tugboat's towing hook. This method provides very good control over the ship, particularly in a fore and aft direction. In many European ports with high tidal ranges, a vessel preparing to berth must first enter and pass through a narrow pair of locks before entering the enclosed harbor basin. It is seen that tug assistance under these circumstances is only possible by means of a head tug towing on a towline attached to the bow of the ship. A stern tug is similarly connected astern to control the ship's stern. Even before entering the lock system, in many cases, the ship is subjected to strong tidal forces and must be controlled by the assisting tugboats.

The head tug, in order to take the ship's headline while underway, must maneuver close under the ship's bow. In this hazardous position, a conventional single screw tugboat must maneuver with greatest care in order to prevent "stemming." After the head tug has taken its position ahead of the ship on the towline, the tug is subjected to the hazard of "girting" (fig. 1). These hazards were thoroughly discussed at the First North American Tug Convention held in Vancouver, B.C., Canada in 1974 (ref. A).

FUNDAMENTAL HAZARDS ASSOCIATED WITH CONVENTIONAL SHIP-HANDLING TUGS

In the United States, tidal action is generally much less, and there are no harbor basins enclosed by a lock system. However, river currents and tidal currents must be accommodated in many United States ports. Traditionally, as in New York City, older finger piers project outwardly at right angles from the shoreline. Lately, terminal construction in the United States has been influenced by the "container revolution," and are now of the "quay" type, with great shoreside container marshalling areas. Oil ports and bulk cargo ports are usually constructed of the "tee" type with the ships laying off the face of the "tee." A system of pile dolphins and catwalks is provided for securing the ship's mooring lines.

SHIP HANDLING METHOD: PORTLAND, OREGON, U.S.A.

Step 1
Positioning

Wind

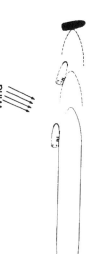

Stemming

Girting

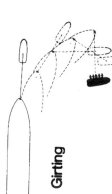

Figure 1

Excerpts from "Ship Handling at Ras Tanura Sea Island" by Philip F. Spaulding (M.Sc.; Life Fellow, S.N.A.M.E; David Taylor Medal, M.A.S.N.E.; President, Nickum & Spaulding Associates, Inc.) presented at the Seventh International Tug Convention and Exhibition, London, England, 15-18 June 1982. Courtesy: Nickum & Spaulding Associates, Inc., Seattle, Washington.

tugs are used to turn the ship, then breast her up to the pier. The ship's athwartship momentum is then checked by the tugs backing down on their headline.

When moving a "dead" ship by the "American" method (fig. 3), a third tug is placed at the stern of the ship opposite the "alongside" tugs. This tug is made up to the ship by the use of three lines. A headline leading through the bow towing eye or around the bow bitt, is lead forward on the ship. A second line is also lead through the same chock tending aft as a spring line. The third line is a stern line leading athwartship directly to the ship's stern.

Properly designed American tugboats have a pronounced shoulder at the forward quarter of the hull, so when making up to handle a "dead" ship, the two forward lines are set up by the use of winches or gypsies. Then, by horsing in the stern line with a stern line towing winch or capstan, the tug pivots on its shoulder setting all lines up violin string tight. This stern tug then provides the power and steering control to move and maneuver the "dead" ship in much the same manner as if the ship had her own power.

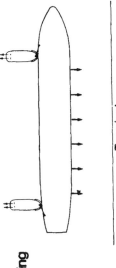

Step 2
Breasting

Terminal

Step 3
Checking or Departing

Terminal

Figure 2

The "American" method, illustrated as that used in Portland, Oregon fig. 2), evolved as the most effective way for assisting ships while they are being maneuvered into finger piers. This is accomplished by using at least two tugboats secured alongside the ship, opposite the side to which the ship would lay to the pier. While in the river or channel with the tugs alongside, the ship's power and steering is used to move the ship, with the tugs giving little control or assistance. The tugs are secured to the ship by the use of their headline leading through a towing eye or around a cruciform bitt located on top of the stem head. While turning into the berth, the

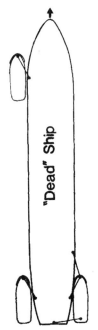

"DEAD" SHIP

"Dead" Ship

Figure 3

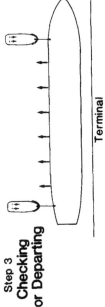

It is noted that positioning tugs alongside, using the "American" method, they are comparatively safe, and the danger of "stemming" and "girting" is minimized. It is not necessary for the forward tug to maneuver under the ship's bow in order to pass its headline to the ship, nor seldom is the tug positioned ahead on a towline, eliminating the danger of "girting."

SHIP-ASSIST TUGBOAT FEATURES AND CHARACTERISTICS

American tugs are usually equipped with a winch or anchor windlass fitted with two gypsies on the fore deck and a towing winch with gypsies or a towing bitt and capstan aft. American tugs are not fitted with towing hooks similar to European tugs, as a quick release of their tow line is not considered paramount. Ship assistance is just one of a multitude of the harbor and coastwise towing duties that American tugs are called upon to perform; therefore, they must be designed for great versatility in their operation. Their deck gear and its location must also reflect these varied requirements.

Europeans have been most active in promoting the tractor tug concept amongst American tugboat operators, and have been very critical of them because of their reluctance to adopt this "modern" design concept. Tractor tugs were developed in Europe to prevent the danger of "stemming" and "girting" which is attendant to the "European" method of ship-assisting, but not so to the "American" method. Whereas it is admitted that tractor tugs have outstanding maneuverability, the issues that have provided the greatest resistance to tractor tug adoption include: first cost and maintenance; being dedicated to ship assistance; as well as their deeper draft and modest towline pull.

At this point, it is interesting to note that the first vertical axis propeller, similar to the Voith-Schneider propeller, was developed by an American. Professor Kirsten, Professor of Aeronautical Engineering at the University of Washington in Seattle, developed this propeller in the early 1920s.

First, cost is always foremost on the minds of American tugboat operators. For this reason, most new tugs are twin screw because of their better maneuverability and off-the-shelf American diesel engine selection. They are fitted with the largest, slowest speed open-water propellers that can be accommodated within the dimensional constraints of the hull and available reverse reduction gear ratios. Some tugs are equipped with triple steering rudders and when fitted to a high, displacement/length ratio hull form, they possess unusual maneuverability.

The next advance in design sophistication is to add four flanking rudders in a fashion similar to that used on the Mississippi River towboats. This feature enhances the tug's maneuverability while it is backing down.

With acknowledging that Dip. Ing. W. Baer of J. M. Voith GmbH, was the father of Thrust Vector Diagrams in a paper he presented at the First Tug Conference, (ref. B), with apologies, I will borrow his thunder. The thrust diagram shown (fig. 4), is used as a basis of comparison with all other tugboat types illustrated and discussed in this paper. The performance of an open-water, propeller-driven twin screw tug in forward thrust, is taken as 100% and all other diagrams are factored therefrom. The dotted lined envelope in figure 4 illustrates the improvement in maneuvering range astern by the addition of flanking rudders.

In cases where American companies have elected to dedicate their vessels to ship assistance, the ultimate designs of American tugboats are known as flanking rudder tugs (FR) of 6000 HP. They are twin screw and fitted with the largest diameter specially-developed fixed propeller nozzles possible, within the di-

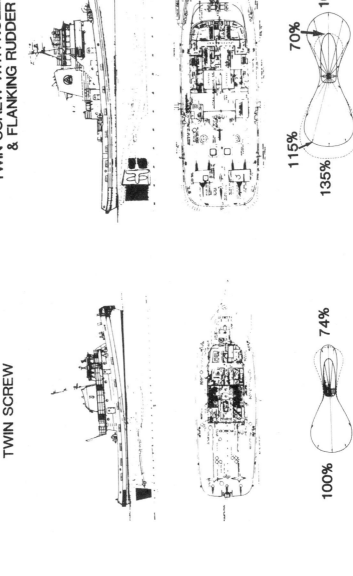

CONVENTIONAL STERN DRIVE
TWIN SCREW WITH NOZZLES
& FLANKING RUDDER

70%   105%

115%

135%

Thrust Diagram

Figure 5

CONVENTIONAL STERN DRIVE:
TWIN SCREW

74%

100%

Thrust Diagram

Figure 4

to push against the bow fender then back away hard on the two headlines. Note the virtual equality between the forward and astern thrust, a most important feature for tugs using the "American" method.

The dotted lined envelope in figure 5 illustrates the thrust derived from a conventional nozzle as compared to the specially developed nozzle and propeller combination used on the lastest flanking rudder tugs.

It is possible, by the proper placement of the steering rudders and flanking rudders, together with rotating the propellers in opposite direction, to physically move the tug athwartship smartly without advancing. Also, the tug can be easily rotated about its axis without forward movement.

mensional constraints of the hull. They are fitted with three steering rudders and four flanking rudders and have a cutaway afterbody to enhance their astern thrust. These flanking rudder tugs (fig. 5), by using the "American" method, perform most of their ship assisting by the bow. They are, however, equipped with a towing hook and capstans on the after deck for an occasional towing assignment.

Two very large headline spooling winches, (fig. 6), are located on the large foredeck of the tug. The winches are equipped with powerful brakes and the two headlines lead forward under a large double cruciform bitt (fig. 7), then to the ship. The tug has an unusually large bow radius and the bow is heavily fendered (fig. 8). The flanking rudder tug's principal ship-assisting maneuver is

CONVENTIONAL STERN DRIVE:
TWIN "Z" DRIVE

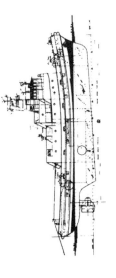

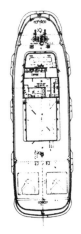

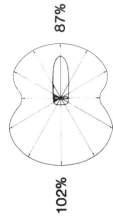

102%   87%

Thrust Diagram

Figure 9

Figure 8

In order to complete the spectrum of ship-assisting tugboats, the following illustrations are included with their trust diagrams. Each of the following tug types incorporates some propeller manufacturer's product, hence, our assessment may be open to argument. However, the illustrations do reflect our experience and unbiased opinion.

The twin propeller "Z" drive with the units located at the stern is typical of Japanese and Korean tugs (fig. 9). Actually, for ship-assisting work, this type of tug configuration acts as a reverse water tractor, in that the vessel works by pushing with the bow or backing down on the headlines in reverse mode.

The astern thrust diagram (fig. 10) was prepared to compare the stern drive propeller "Z" drive with water tractor types. The illustration clearly shows that the aft drive permits designing a vessel of minimum draft, however, astern bollard pull is reduced by propeller wash impingement on the hull itself (thrust deduction).

Two basic types of water tractor tugs are shown. The water tractor with twin propeller "Z" drive (ref. C), with its thrust diagram, (fig. 11), can be compared with a water tractor tug with twin Voith-Schneider (VSP) drive (ref. B), and its thrust diagram (fig. 12). The features of tractor tugs have been broadly discussed at previous tug conferences, however, salient features and advantages are repeated here in order to relate to the assistance of large tankers while berthing at Port Hypothetical:

1. Steering propulsion units are located forward of the "hook" eliminating the risk of "girting", (fig. 1).

2. With the steering propulsion units located forward, after taking the ship's bowline, the tug can pull away from the vessel's bow, eliminating the danger of "stemming," (fig. 13).

3. The forward position of the steering propulsion units minimizes the risk of fouling the towline.

4. Tractor tugs push with their sterns when the propulsion units are in the astern mode. If they want to take a strain on their towline, the propulsion units are quickly placed in a head mode without a change in direction of the tug.

5. The Voith-Schneider (VSP) tractor tug can change its direction and thrust by simply adjusting the angle of attack of their rotating blades with no change in engine speed, hence, they have the quickest response time. They steer with equal ease in any direction — backward, forward or sideways.

ASTERN THRUST DIAGRAM
ASTERN DRIVE TUG VS. WATER TRACTOR TUG

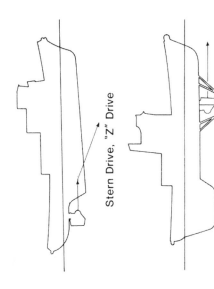

Stern Drive, "Z" Drive

Water Tractor, "Z" Drive

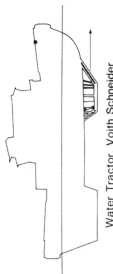

Water Tractor, Voith Schneider

Figure 10

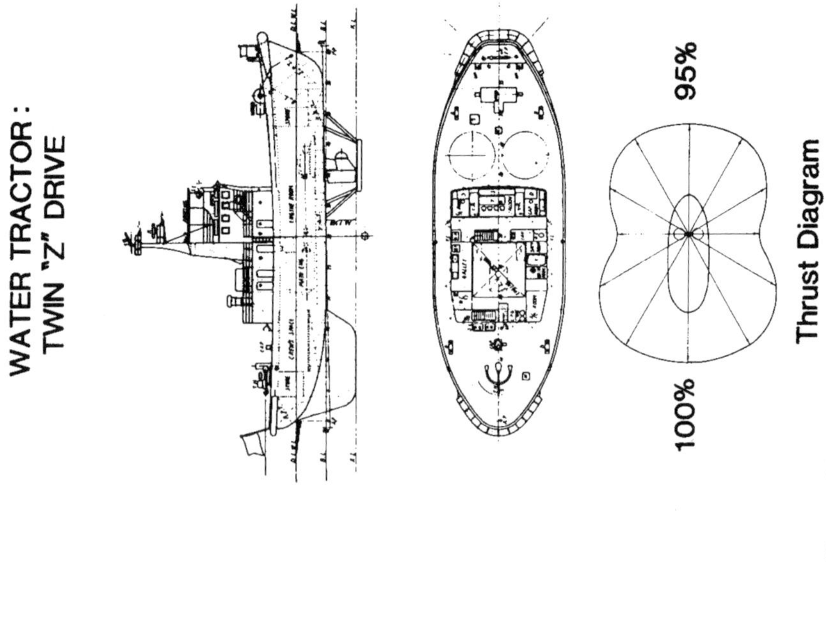

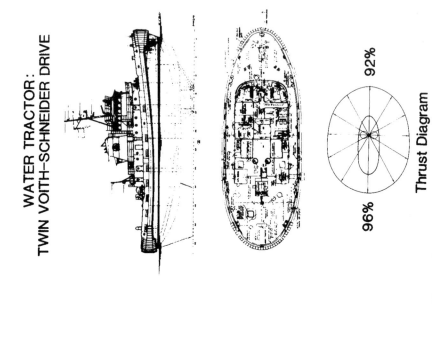

WATER TRACTOR :
TWIN VOITH-SCHNEIDER DRIVE

92%

96%

Thrust Diagram

Figure 12

WATER TRACTOR :
TWIN "Z" DRIVE

95%

100%

Thrust Diagram

Figure 11

TUGBOAT THRUST

PROPULSION TYPE	BOLLARD PULL #/ SHP	
	AHEAD	ASTERN
TWIN SCREW W/ OPEN PROPELLERS	25.9	19.4
TWIN SCREW W/ OPEN PROPS, & FLANK. RUD.	25.8	19.1
TWIN SCREW W/ CONVENTIONAL NOZZLES	34.9	18.1
TWIN SCREW/SPEC. NOZZLES & FLANK. RUD.	29.7	27.2
STERN PROPELLER "Z" DRIVE	26.4	22.5
WATER TRACT. TWIN PROP. "Z" DRIVE	25.9	24.8
WATER TRACT, TWIN VSP DRIVE	24.9	23.7

Service Factor 10%

Figure 14

SHIP-ASSIST TUGBOAT OPERATING PHILOSOPHY

Dispatch and safe operation in berthing large ships must be foremost when rationalizing the parameters for a ship-assisting tugboat system. Dispatch is essential when one considers the hourly cost of operating large tankers or container ships with their valuable cargoes. Safe operation is paramount, as damage to the vessel or pier can be astronomical in cost, to say nothing about the casualty's affect on the environment.

When observing the berthing operations at the oil port of Valdez, Alaska (two million barrels of oil per day); Ras Tanura, Saudi Arabia (ten million barrels of oil per day); and the container port of Seattle (second largest in the United States) one thing must be borne in mind. Procedures for the berthing operation are discretionary with the pilot and ship's master, and not all pilots

TAKING THE BOWLINE

(NO DANGER OF STEMMING)

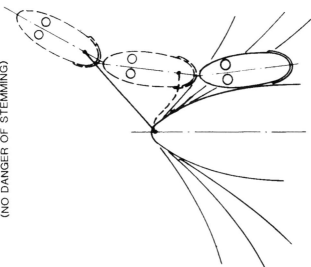

Figure 13

Figure 14 summarizes the bollard pull of the various types of ship-assisting tugs in pounds of bollard/SHP .

approach a problem in exactly the same manner. For this reason, any plan for developing a tugboat ship-assisting system must be reviewed and approved by the harbor pilots.

Because different ports vary greatly in the number of terminals, together with vastly different number of ship arrivals and departures requiring tug assistance, this paper's analysis is confined to those tugs required for each ship assist. As Port Hypothetical is to be an oil port, ship-assisting procedures are to follow the general practice already established through successful operating experience.

OIL PORT SHIP ASSISTING PROCEDURE

In this procedure, usually two, three or more tugs are positioned and secured on the offshore side of the ship in such a way as to produce the thrust control, primarily for the breasting movement of the vessel. Fore and aft thrust as well as minor steering control is produced, or at least augmented, by the ship's propeller and rudder. The ship normally will be brought "dead" in the water, parallel to the berth a short distance out. The ship will then be breasted up to the face of the berth with the tugs supplying the lateral thrust necessary. At the proper time, at the pilot's discretion, the tugs will back down applying the necessary thrust to check the vessel's athwartship motion (fig. 2).

It will be shown herein that a certain aggregate thrust is required in the tug flotilla being used to berth a ship of a given mass under the most unfavorable conditions anticipated at a given location. Obviously, this aggregate thrust can be obtained with a small number of very powerful tugs or a greater number of less powerful tugs. The upper limit of tugboat power is directly related to the design of the fendering system so as to prevent damage to the ship's hull structure. All things being proportional, the fewer the tugs, the less the cost; and because of communication, it is easier with fewer tugs for the pilot to manage the berthing operation.

As a typical example, when making a study for a specific port's ship-assisting tugboat system, we dealt with the Marine Department and Terminal Pilots in Ras Tanura, Saudi Arabia. They considered that the largest suitable tug was a twin screw flanking rudder tug producing 6000 HP with an overall length limitation of 136 feet, (fig. 5).

At Ras Tanura, where they handle the largest vessels in the world on a regular basis, they have developed an advanced phase of the "European" method for ship assistance. The head tug and stern tug are both water tractor Voith-Schneider-propelled (VSP) tugs of 4000 HP each and are used for positioning the tug abreast the berth (fig. 12). Amidship, depending on the weather, one or two flanking rudder (FR) tugs of 6000 HP each are used for the breasting operation, and they then check the athwartship movement as the vessel approaches the pier, (fig. 15). When departing, a fully loaded tanker requires less critical tug assistance, as the operation does not involve bringing a large mass to rest against a stationary object. However, the mass of the departing loaded tanker will be more than twice that of the ballasted tanker coming in.

SHIP HANDLING METHOD: RAS TANURA, SAUDI ARABIA

Step 1
Positioning

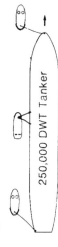

250,000 DWT Tanker

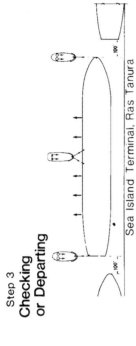

Step 3
**Checking
or Departing**

Sea Island Terminal, Ras Tanura

Figure 15

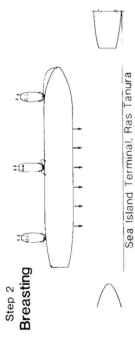

Step 2
Breasting

Sea Island Terminal, Ras Tanura

REFERENCES

Armstrong, Macolm C., Pilot. 1980. *Practical Shiphandling*, First Edition. Glasgow: Brown, Son and Ferguson, Ltd.

Brooks, Gregory Captain, and Captain S. Wallace Slough. Escorting Ships with Tractor Tugs. August/September 2000. *Professional Mariner* 50.

Cordage Institute *International Guideline CI 1401-06 Safer Use of Fiber Rope*. May 2006.

Crowley Maritime Services Inc. vs. Maritrans Inc., Collision between the Tug Sea Kink & the Tanker Allegiance, 04-35724, United States Court of Appeals for the Ninth Circuit, 2002.

Crowley, Michael. November 2007. Bender Fenders. *WorkBoat Magazine*.

Gray, David LPE, and PE Bruce L. Hutchison. Prepared for the State of Washington: Department of Ecology. 2004. *Study of Tug Escorts in Puget Sound*. File number 04075.

Gray, David LPE, and EIT Elizabeth Reynolds. *Engineering Methodologies Used in the Preparation of Escort Tug Requirements for the Ports of San Francisco and Los Angeles/ Long Beach*. Presented at the Society of Naval Architects and Mariner Engineers (SNAME) California Joint Sections Meeting, Monterey, CA, May 13-15, 1999.

Griffin, Barry, and Blaine Dempke. July 2006. High Performance Winches. *Pacific Maritime Magazine*.

Griffin, Barry. July 2000. High Performance Ropes: New Strength, Silent Stress. *Pacific Maritime Magazine*.

Hensen, Henk, FNI. 2006. *Bow Tug Operations*. London: The Nautical Institute.

Hensen, Henk, FNI. 2003. *Tug Use in Port, Second Edition*. London: The Nautical Institute.

Livingston, George H, MNI, and H. Grant, MNI. 2006. *Tug Use Offshore in Bays and Rivers*. London: The Nautical Institute,

Mott, Harvey J., Dirk H. Kristensen, Duane H. Laible, Anthony Thompson, and Mike Stone. *Acquisition of Tanker Escort and Assist Tugs for Newfoundland Transshipment LTD.*

NautiCAN. September 20, 2003. Bollard Pull Test Results and Comparisons. *Maritime News*.

Reid, George H. 2004. *Primer of Towing, Third Edition*. Centerville, Maryland: Cornell Maritime Press, Inc.

Puget Sound Harbor Safety Committee. 2003, 2005. *Puget Sound Harbor Safety Plans*.

Puget Sound Rope. 2007. *Rope Care and Usage.*,

Roberts, Phil, Danielle Stenvers, Paul Smeets, and Martin Vlasboom. 2002. Residual Strength Testing of Dyneema Fiber Tuglines. Paper presented at the International Tug and Salvage Convention.

Rowe, R.W., FNI. 2004. *The Shiphandlers Guide, Second Edition*. London: The Nautical Institute.

Reid, George H. 2004. *Primer of Towing, Third Edition*. Centerville, Maryland: Cornell Maritime Press, Inc.

Reid, George H. 1986. *Shiphandling with Tugs, First Edition*. Centerville, Maryland: Cornell Maritime Press, Inc.

Rowe, R.W., FNI. 2004. *The Shiphandlers Guide, Second Edition*. London: The Nautical Institute.

Samson Rope Technologies, Inc. 2005. *Comparison of Fiber Characteristics. Samson Technical Bulletin.*

— 2005. Rope Inspection and Retirement. Samson Technical Bulletin.

Schisler V. J. Captain, and Captain Gregory Brooks. Team Towing Using Relatively Small Tractors on Heavy Ships. Towing Solutions, Inc. http://www.towingsolutionsinc. com/technology-tractor_tandem.html (accessed January 2008).

Voith Turbo Schneider Propulsion GmbH & Co. *Voith Water Tractor Manoeuvre Manual.* http://www.voithturbo.com/applications/documents/document_files/666_e_am_vwt_manoeuvre_manual_e.pdf (accessed January 2008).

INDEX

ABOUT THE AUTHOR

Captain Jeff Slesinger, born in 1953, has had two seagoing passions. One is the challenge of navigating and piloting a variety of vessels in their assigned trades; the other is training and teaching those who wish to acquire the skills of a professional mariner. His seagoing career began at an early age on sailing vessels, but he has devoted the past 30 years to working on tugs and the barges and ships they tow. A licensed master since 1977, he has embarked on voyages from harbors in New England throughout the Caribbean, Pacific Coast and to the Bering Sea in Alaska.

Captain Slesinger has created onboard and simulator training programs for tug handling, bridge resource management and other skills critical to a professional mariner. He continues to work with training institutions and individual companies in the development of programs that foster the acquisition of practical expertise required to be competent and qualified at sea.

His current principal activity is presiding over his company, Delphi Maritime, LLC which is devoted to creating innovative solutions for working safely at sea. However, Captain Slesinger continues to operate tugs in the Pacific Northwest as a master and onboard instructor.

ISBN-13: 978-0-87033-598-3

9 780870 335983

54000